IEE CIRCUITS, DEVICES AND SYSTEMS SERIES 16

Series Editors: Dr R. Soin
Dr D. Haigh
Dr J. Wood
Professor Y. Sun

Wireless communication circuits and systems

Other volumes in the Circuits, Devices and Systems series:

Wireless communication circuits and systems

Edited by
Yichuang Sun

The Institution of Electrical Engineers

Published by: The Institution of Electrical Engineers, London,
United Kingdom

© 2004: The Institution of Electrical Engineers

The Institution of Electrical Engineers,
Michael Faraday House,
Six Hills Way, Stevenage,
Herts., SG1 2AY, United Kingdom

www.iee.org

British Library Cataloguing in Publication Data

Wireless communication circuits and systems. – (IEE circuits,
   devices and systems series ; 16)
1. Wireless communication systems 2. Electronic circuits
I. Sun, Yichuang II. Institution of Electrical Engineers
621.3′82

ISBN 0 85296 443 9

Typeset in India by Newgen Imaging Systems, Chennai
Printed in the UK by MPG Books Limited, Bodmin, Cornwall

To Xiaohui, Bo and Lucy

Contents

Contributors

David J. Allstot
Department of Electrical Engineering
University of Washington
Seattle, WA 98195-2500
USA

Kiyong Choi
Department of Electrical Engineering
University of Washington
Seattle, WA 98195-2500
USA

Mourad N. El-Gamal
Department of Electrical and Computer Engineering
McGill University
3480 University Street
Montreal, P.Q. H3A 2A7
Canada

Mohamed M. Hafed
DFT MicroSystems, Inc.
460 St. Catherine Street West, Suite 312
Montreal, Quebec H3B 1A7
Canada

Mona M. Hella
RF Micro Devices
296 Concord Road
Billerica, MA 01821
USA

John Hughes
Philips Research Laboratories
Cross Oak Lane
Redhill
Surrey RH1 5HA
United Kingdom

Mohammed Ismail
Department of Electrical Engineering
The Ohio State University
Columbus, OH 43210
USA

Rushikesh Lala
Department of Electronic, Communication and Electrical Engineering
University of Hertfordshire
Hatfield
Herts AL1O 9AB
United Kingdom

Brian J. Minnis
Philips Research Laboratories
Cross Oak Lane
Redhill
Surrey RH1 5HA
United Kingdom

Paul A. Moore
Philips Research Laboratories
Cross Oak Lane
Redhill
Surrey RH1 5HA
United Kingdom

James Moritz
Department of Electronic, Communication and Electrical Engineering
University of Hertfordshire
Hatfield
Herts AL1O 9AB
United Kingdom

Jinho Park
Department of Electrical Engineering
University of Washington
Seattle, WA 98195-2500
USA

Arun Ravindran
Department of Electrical Engineering
The Ohio State University
Columbus, OH 43210
USA

Gordon W. Roberts
Department of Electrical and Computer Engineering
McGill University
3480 University Street
Montreal, P.Q. H3A 2A7
Canada

Anup Savla
Department of Electrical Engineering
The Ohio State University
Columbus, OH 43210
USA

Rungsimant Sitdhikorn
Department of Electronic Engineering
Mahanakorn University of Technology
Bangkok 10530
Thailand

Adrian Spencer
Philips Research Laboratories
Cross Oak Lane
Redhill
Surrey RH1 5HA
United Kingdom

Yichuang Sun
Department of Electronic, Communication and Electrical Engineering
University of Hertfordshire
Hatfield
Herts AL1O 9AB
United Kingdom

Tommy K. Tsang
Department of Electrical and Computer Engineering
McGill University
3480 University Street
Montreal, P.Q. H3A 2A7
Canada

Apisak Worapishet
Mahanakorn Microelectronics Research Centre (MMRC)
Mahanakorn University of Technology
S1/1 Cheumsampan Road
Bangkok 10530
Thailand

Seoung-Jae Yoo
Department of Electrical Engineering
The Ohio State University
Columbus, OH 43210
USA

Md. Iqbal Younus
Department of Electrical Engineering
The Ohio State University
Columbus, OH 43210
USA

Preface

This book is concerned with integrated circuits and systems for wireless and mobile communications. Circuit techniques and implementation of reconfigurable low-voltage and low-power single-chip CMOS transceivers for multiband and multi-mode universal wireless communications are the focus of the book. Applications encompass both long-range mobile cellular communications (GSM and UMTS) and short-range wireless LANs (IEEE802.11 and Bluetooth). Recent advances in research into transceiver architecture, RF frontend, analogue baseband, RF CAD and automatic testing are reported. The book, containing ten chapters, covers topics including receiver architectures, reconfigurable and programmable analogue base-band arrays, design and on-chip tuning of active IC filters including polyphase filters, receiver front-end modules (LNA, VCO, mixer) and chips, transmit power amplifiers, parasitic-aware RF design and optimisation, and on-chip testing of communication ICs. The book can be used as a text or reference book for a broad range of readers from both academia and industry.

I would like to thank Robin Mellors-Bourne, Director of the Publishing Department of the IEE for his support and encouragement. My gratitude also goes to David Haigh of UCL, IEE Circuits, Devices and Systems Series Editor, for his help, in particular with Chapter 1. Assistance from the editorial and production team at IEE is acknowledged. I am very grateful to the authors of the chapters for their great effort and high speed in contributing such high-quality chapters. Their professionalism is greatly appreciated. I should also thank my lovely family, Xiaohui, Bo and Lucy for their understanding and support.

During the preparation of this book, I was also Guest Editor of a special issue on 'RF circuits and systems for wireless communications' for *IEE Proceedings: Circuits, Devices and Systems* (published in December 2002). Some of the book chapter contributors have written papers for that special issue. I am pleased to see that some of the book chapters have benefited from the special issue.

Yichuang Sun

Chapter 1

Wireless communications and transceivers: an overview

Yichuang Sun

1.1 Wireless and mobile communications

Wireless and mobile communications is one of the fastest growing areas of modern life. It has an enormous impact on almost every aspect of our daily lives. A world of ubiquitous wireless devices is emerging, from wireless sensors and tags to mobile terminals. There are a number of wireless systems and networks in operation such as GSM, GPRS, EDGE, IS-95, UMTS, IMT2000, DECT, IEEE 802.11, and Bluetooth, and several multiple access techniques including MC-CDMA, W-CDMA, and OFDM are used. Future-generation wireless communication, networking and connectivity beyond the so-called 3G have already attracted international attention.

Mobile cellular communications, wireless LANs and broadband wireless access are the three most popular application areas. Long-range cellular communication encompasses the evolution of GSM (2G+: GPRS, EDGE), IS-95 and 3G systems (IMT2000/UMTS). It aims to achieve full mobility and global roaming for individual links maximising system capacity. In short-range wireless networking and wireless connectivity, WLAN or PAN covers local hot spot areas for very high data rate asymmetric individual links (e.g. IEEE 802.11, HIPERLAN), and universal wireless networking and home connectivity are now achieved by Bluetooth and DECT. Fixed wireless local loops provide wireless access from street kerb to home for broadband internet and multimedia services. Space-time coding and ultra-wideband modulation have stood out particularly as the hope for future wireless and mobile communication systems.

The future wireless and mobile communication system will be universal, suitable for different wireless networks such as the W-CDMA-based cellular mobile system, OFDM-based wireless LANs and inexpensive Bluetooth. The mobile terminal or transceiver in such a universal system must be reconfigurable to meet multiband and multimode requirements and must be of small size and low power to drive the cost down. This book concerns itself with this important subject.

1.2   Integrated circuit technologies for wireless systems

Low-cost miniaturised transceivers can only be achieved by advanced IC technologies. Today's microelectronics is moving towards SoC (system on a chip) implementation for communication and information systems. There have been several competing IC technologies, which include CMOS, BiCMOS, Bipolar, SiGe and GaAs. Low-cost and readily available CMOS has to date been the mainstream technology used in many areas, including computers, video, image and display systems, DSL and cable modems and circuits for wired communications, RF circuits and transceivers for wireless communications, and even high-speed circuits and systems for optical communications. Recently CMOS has been demonstrated to be a viable technology for very high bit rate broadband circuit design at over 10 Gb/s in highly integrated systems. Meanwhile advances in device scaling and doping profile optimisation have also resulted in SiGe bipolar transistors with impressive performance, including cut-off frequencies in the region of 200 GHz. This makes them an attractive choice for applications at 40 Gb/s and above.

Modern microelectronic circuits and systems are mainly based on submicron- and deep-submicron-CMOS technologies. Whilst pushing the limits of CMOS, preparation for the post-CMOS era is well under way. Many other potential alternatives such as MEMS have been pursued. Nanoelectronics and nanotechnologies have recently emerged and have been widely investigated. They will certainly become the key enabling technologies in future advances. With the current research trend, nano-CMOS technology may soon be available for integrated communication circuits and systems.

In this book, CMOS technology is used throughout for implementation of transceiver, analogue and RF circuits for wireless communications, because of the promising RF performance of submicron-CMOS, together with the advantages of low cost and ease of integration with baseband digital circuitry for SoC.

1.3   Transceivers for wireless and mobile communications

As discussed above, transceivers play a key role in the successful applications of wireless communication, networking, accessing and connecting. Apart from these well-known application areas, transceivers have also been widely used in many other applications such as backplane, ambient intelligence, and embedded systems. Actually, transceivers will become an essential device in any future intelligent electronic system as wireless connectivity becomes more and more popular.

Today's transceivers are tending towards single-chip CMOS devices with very small physical size, very low power consumption and very low cost. SoC transceivers involve a mixed-signal design containing both analogue and digital circuits. Integration of the RF frontend and channel filters of a transceiver is particularly challenging. Although great progress has been made, real single-chip transceivers having everything on one chip are not commonplace.

Recently, multimode transceivers for Bluetooth and 802.11, or GSM, IMT2000 and UMTS have been receiving particular attention. Now, universal transceivers for all types of application including local networking, home connectivity, and cellular communications are also being pursued. Reconfigurability, programmability and tunability are a must in such transceivers. In these transceivers, the operating frequency may be variable and the channel bandwidth may be variable to accommodate different mobile and wireless standards. The carrier frequency, for example, is typically within the range 1–5 GHz. The major design challenges over variable carrier frequencies and bandwidths are transceiver filtering and linearity. Future transceiver design must also consider requirements of smart antennas and ultra-wideband radio.

Software and DSP-based concepts have been hot topics in so-called software-defined radio (SDR). There are some very tough challenges for use of SDR in high-performance long-range wireless communications. Firstly, the spectrum allocated to 3G is above 2 GHz. At such carrier frequencies the pure software radio approach will for some time remain impractical, since analogue-to-digital conversion at such carrier frequencies remains some way in the future. Secondly, the complexity of the 3G air interfaces requires significantly greater processing than second-generation systems. This precludes low-power terminals using traditional DSP devices. Realistically, direct conversion at the antenna in handsets will not be feasible for many years. Even for GSM standards there is still some way to go towards pure software radio. On the other hand, for low performance requirements in short-range wireless connectivity applications such as Bluetooth, complex DSP techniques may not be a high priority. In such applications, low cost, low power, and small-size implementation is more important; therefore RF IC may be more crucial than DSP. The fact is that no matter how fast DSP is, the power consumption tends to be simply too high for many portable applications. How far DSP can be pushed towards the antenna is not just up to the DSP technology itself, but is also dependent on advances in the analogue part of the system including the analogue-to-digital converter (ADC). There is a trade-off between high performance multifunctionality and single-chip integratability for transceivers. In this book we try to obtain practical solutions for optimum transceiver design in the RF and analogue domains.

1.4   Challenging topics in communication circuits and systems

In this section we highlight some challenging topics in analogue and RF CMOS circuits and systems for wireless communications. These topics are critical factors for the implementation of future-generation ubiquitous single-chip transceivers.

1.4.1   *Transceiver architecture and system-level design*

There are three types of receiver architecture: heterodyne, low-IF, and zero-IF. Traditional receivers normally use the heterodyne architecture with one or two high-IF stages. Software-defined radio, however, drives the use of zero-IF receivers. The heterodyne receiver is difficult for single-chip implementation, as it requires a sharp image reject filter. The difficulties with the zero-IF structure are the DC offsets, $1/f$ flicker noise and second-order distortion, and the very stringent requirements for data converters. Currently a compromised low-IF architecture is most widely used. Using this type of receiver, DC offsets can be overcome and high integration achieved. The quadrature IF or baseband architecture with orthogonal I- and Q-channels is used in receivers. On the transmitter side, there is a similar development in architecture.

In the system-level design of a transceiver, the location of the data converter in the receiver chain is important for the whole system performance. In some software-defined radio systems, the ADC is ideally placed at the antenna. However, very large dynamic range and very high-speed converters are required and there are technological limitations on achieving this. Thus the choice between certain analogue and digital signal processing techniques must be made realistically to meet all requirements of power consumption, chip area, and reconfigurability. Several methods have been proposed to relax the requirements on ADCs, such as gain control using a variable gain amplifier (VGA) to accommodate the fixed input range of an ADC. More recently a revised low-IF receiver architecture in which the need for a complex ADC is avoided has been proposed. In this architecture, only the real part of the complex IF signal after mixers is processed, the imaginary part being ignored. The new architecture substantially simplifies the ADC design whilst retaining all the advantages of a digitised, low-IF receiver. This architectural and system level topic will be covered in Chapter 2 of the book.

1.4.2   RF, IF and analogue baseband circuits and subsystems

As discussed above, system design issues of CMOS transceiver hardware architecture should be addressed very carefully to achieve the best compromise between high programmability, low power consumption, and small chip area. A transceiver consists of several parts including a digital baseband, data converters, an analogue baseband, an IF stage, and an RF frontend. Digital baseband design involves DSP software and FPGA hardware implementations. The major advantage of DSP and FPGA is the ease of reconfigurability and programmability. Current research activities in digital baseband design are directed to further enhance computation speed and reduce power consumption. For ADC design, a key trade-off is between bandwidth and noise performance; higher bandwidth will allow more noise to enter the ADC. Recently, bandpass $\Sigma\Delta$ modulation techniques have been proved useful for reducing the noise in a wideband ADC.

To bring a whole transceiver onto a single chip, the key components are, however, the mixers, oscillators and filters and the LNAs in the receiver frontend and power amplifiers in the transmitter. In the baseband and IF stages the channel select filter and VGA are important. Generally, the major advantages of analogue signal processing are high speed, low power consumption and small chip area. Reconfigurability and tunability of analogue and RF circuits can also be achieved using various recent circuit techniques. Chapters 3–8 in this book will therefore focus on the analogue baseband, the IF stage and RF frontend circuits and subsystems. A review of the topics covered in these chapters is given below.

1.4.2.1 Reconfigurable analogue circuits and FPAA for the baseband and IF stages

Reconfigurable and programmable analogue circuits and systems have received much interest recently. Op-amp-based active-RC and MOSFET-C circuits have been used in transceiver baseband design and OTA-based OTA-C circuits have also been used in IF stages. Both MOSFET-C and OTA-C circuits are electronically tuneable. With some additional digitally controlled switches, the baseband and IF chains can also be made reconfigurable for multiband and multimode wireless applications. The tuneable frequency and variable gain chain can also relax the stringent requirements of speed and dynamic range on ADCs. The emerging field-programmable analogue array (FPAA) may make it possible for a whole baseband and IF stage to be fully programmable and reconfigurable. The FPAA for this purpose can be constructed using OTA-C and MOSFET-C circuit techniques. With the successful application of digitally controlled analogue baseband chains, the use of more general programmable analogue and mixed-signal arrays in multistandard transceivers, especially in analogue baseband and IF stages typically containing lowpass filters and variable gain amplifiers, seems possible in the near future. A whole programmable mixed-signal baseband array to contain the analogue FPAA, analogue-to-digital converter and digital FPGA may also be possible.

1.4.2.2 Filter design and on-chip tuning methods for applications up to RF

Filters are widely considered as the bottle-neck for ubiquitous system-on-chip communication due to the difficulties in implementing them on-chip. The use of narrowband RF filters for channel selection and image rejection has been delayed due to lack of efficient monolithic fabrication. Most of the actual systems use off-chip bulky passive filters. Major problems for active IC filters are lack of high-quality inductors, good varactors and efficient automatic tuning systems. In low-IF and zero-IF transceivers, complex polyphase filters are required. Recently, much attention has been paid to the design and implementation of this type of filter. To satisfy all transceiver design requirements, many different types of filters operating over a wide range of frequencies and bandwidths are required. Popular ones are active-LC filters, OTA-C filters and MOSFET-C filters for RF, IF and baseband applications, respectively. Different filter circuits have different design challenges. For example, optimisation of LC tanks minimises the effects of loss and parasitics of on-chip inductors for active-LC filters and design of highly linear OTAs is needed for OTA-C filters. Various parasitics and large variations in on-chip component values (which can be as high as 50 per cent) lead directly to large divergence of the achieved filter response from the intended design specifications. On-chip tuning thus becomes essential and the tuning system may well be the most difficult challenge in achieving satisfactory filter performance.

1.4.2.3   RF receiver frontend modules (LNA, VCO, mixer) and chips

Low noise must be targeted in RF receiver frontend design, as the received signal is normally very weak. The high carrier frequency is another feature of the frontend. Wireless communication covers the 1–2 GHz mobile cellular range to the 5 GHz range for WLANs. Traditionally, radio-frequency integrated circuits (RFICs) were implemented in GaAs or SiGe bipolar technologies, because of their relatively high unity gain cutoff frequencies f_T (>65 GHz) and their superior noise performance. However, as the minimum feature size of CMOS devices decreases, the f_T of the transistors continues to improve to the point where it is becoming comparable to those of GaAs and SiGe processes. Deep-submicron-CMOS devices with f_T's exceeding 100 GHz and minimum noise figures (NF) less than 0.5 dB at 2 GHz have been possible. Because of these promising RF performances, together with the advantages of low cost and ease of integration with baseband digital circuitry, CMOS is becoming a viable alternative for RF applications. Recent effort has been directed to integrated inductors riding on rapid advances in IC technology. LC tanks and transformers in CMOS and even transmission lines are now possible and have performances suitable for many applications, especially short-range wireless connectivity. The success of integrating inductors on silicon chips has led to a major renewal of RF research activity. Various traditional RF circuits based on discrete inductors have been redesigned to benefit from on-chip inductors. In particular, transceiver frontend circuits using LC tanks such as low-noise amplifiers, oscillators, mixers and filters can now be integrated on the same chip.

1.4.2.4   RF transmit power amplifiers

Power amplifiers in the transmitter are typically working at high signal level and high carrier frequencies. Until now, power amplifiers for wireless applications were produced almost exclusively in GaAs technologies, with a few exceptions in LDMOS, Si BJT, and SiGe HBT. While CMOS provides high functionality and complexity at low cost, for an RF power amplifier, the problem of using CMOS technology is more severe than other blocks in the transceiver due to the limited voltage-handling capability. The linearity and power efficiency are lower than other technologies. Therefore, implementation of RF power amplifiers for wireless transmitters in CMOS has been one of the most challenging tasks. The design of power amplifiers in CMOS technology is affected by many factors such as low breakdown voltage of deep-submicron technologies. Several classes of power amplifiers may be chosen for different applications. On-chip power controllability or programmability is particularly important for both single- and multistandard wireless applications. The output impedance matching network may be implemented either on-chip using silicon inductors or more easily off-chip as part of the antenna system.

1.4.3   *Design optimisation and automatic testing techniques*

Tools and methods are also important for successful optimum design and on-chip testing of transceivers and RF circuits. Advances in these areas are briefly overviewed below. These topics will be the subjects of Chapters 9 and 10 of the book.

1.4.3.1   RF design optimisation and CAD

Traditional RF design is mainly based on some empirical, trial and error methods. Complex modern integrated RF IC design can only be solved using electronic design automation tools. Two major problems in the design and optimisation of RF frontend circuits in fine-line CMOS technology are active devices that are inferior to their GaAs and SiGe counterparts, and low-quality parasitic-laden passive components owing to the use of the lossy silicon substrates. To overcome these drawbacks, the parasitic-aware synthesis paradigm has been developed. Simulated annealing is one of the key algorithms used in the parasitic-aware design and optimisation methodology. Unlike in baseband circuits, the parasitic effects are severe in RF circuits. In the case of baseband circuit design, minimizing parasitics is usually sufficient. However, the parasitics must be carefully modelled and considered as part of the design process in high-frequency circuit synthesis. If parasitics are considered as part of the design, it is difficult to find an analytic solution by hand. Thus, computer-aided RF design and optimisation must be adopted.

1.4.3.2   On-chip testing of communication ICs

There has been some growing interest in the testing of communication system-on-chip devices including analogue, mixed-signal and RF circuits. Automatic testing becomes more and more important to drive down the overall cost of communication devices due to the decidedly imperfect nature of the manufacturing process and its associated tolerances. Design for testability (DFT) has been widely accepted by industry. In today's IC design, testing issues are considered at the early design stage. Built-in-self-testing (BIST) has already been widely used in many practical electronic products and the on-chip test system concept has also been proposed for complex integrated systems. Both functional testing and parametric testing are important, with functional testing being relatively easier than parametric testing. For analogue, mixed-signal and RF circuits, testing proves much more difficult than for digital circuits due to tolerances, parasitics and non-linearities, which make testing ambiguous and computationally intensive. Analogue circuits consist of many parameters and are defined by a large set of specifications. Attempting to verify all specifications and test all parameters in the production test phase is too prohibitive, so careful planning of which specifications and parameters to test is important. To make the test results unambiguous, care must be taken on test-point selection, test-signal generation and measurement. A recent trend in test integration which promises to reduce many of the burdens of analogue testing is the use of embedded mixed-signal test cores, which are integrated circuit 'macros' that emulate the functions of automatic test equipment. These embedded test cores are designed to perform DC curve-tracing, oscilloscope, timing, and frequency domain measurements using compact and mostly digital integrated electronics.

1.5   Summary of chapters

This book is concerned with CMOS integrated circuits, systems and transceivers for wireless and mobile communications. It contains ten chapters and covers all the key challenging topics highlighted in Section 1.4. The material and chapters in the book are organised around three parts: transceiver architecture and system level issues, constituent components and subsystems, and RF optimisation and testing, as already mentioned in Section 1.4. Transceiver architecture and system level design are first dealt with in Chapter 2, with a novel receiver architecture being detailed. In Chapters 3–8, the building blocks and subsystems in transceivers are then discussed in order from analogue baseband to RF frontend. In more detail, Chapters 3 and 4 are concerned with reconfigurable and programmable analogue baseband arrays, with Chapter 3 giving a digitally controlled analogue baseband chain and Chapter 4 dealing with general programmable analogue array. Design and on-chip tuning of active IC filters are then introduced in Chapters 5 and 6, with Chapter 5 presenting a detailed design of a polyphase channel filter for a low-IF stage and Chapter 6 giving a more general coverage of filter design and tuning techniques for applications up to RF. Further in this part, Chapter 7 addresses the receiver RF frontend modules (LNA, VCO, mixer) and chips, and Chapter 8 provides a tutorial on transmit RF power amplifiers. Chapters 9 and 10 form the third part of the book, dealing with parasitic-aware RF design and optimisation (with power amplifier design examples), and on-chip testing of communication ICs, respectively.

Chapter 2

Non-complex signal processing in a low-IF receiver

Brian J. Minnis and Paul A. Moore

2.1 Introduction

Cellular radio telephone handsets are now in common use throughout the world. The resultant commercial pressure for compact and low-cost products has led to renewed global research interest in receivers and the consequent development of highly integrated solutions based on architectures that might otherwise have languished as mere curiosities. The resultant widespread acceptance of zero-IF (intermediate frequency) and low-IF receivers in handsets for GSM (Global System for Mobile communications) has taken integration to a level from which any further significant improvement is now hard to imagine. But the additional need to address new, third-generation (3G) standards, such as UMTS (Universal Mobile Telecommunication System), is tending to re-focus research activities in this area towards increased flexibility and the maximisation of hardware re-use. This is most easily addressed by increasing the level of digitisation. However, unless this is handled intelligently, power consumption can easily be increased over and above that used by existing analogue implementations, with unacceptable consequences in terms of battery life. Hence, we present a novel architectural solution to this problem, mainly in the context of GSM but which forms part of a multimode receiver for both GSM and UMTS.

After describing the evolution of integrated receiver architectures that can be applied to GSM, this chapter will discuss the digitisation problem in a little more detail, examine the basic performance requirements and derive some of the most important receiver dimensions. It will then present a revised low-IF receiver architecture in which the need for a complex analogue-to-digital converter (ADC) is avoided. The technique involves dropping the Q component of the IF output of the frontend and then only processing the I component in the ADC. Details will be given of the three blocks most closely associated with the non-complex signal processing, namely the polyphase image-rejection filter, the digital make-complex filter and the ADC itself which is based on a fifth-order sigma-delta ($\Sigma\Delta$) modulator. System simulations will be described and the results derived from them will be presented in validation of the modified architecture. The chapter will then discuss the incorporation of the new approach into the multimode receiver for GSM and UMTS before drawing some final conclusions.

2.2   Evolution of integrated receiver architectures

2.2.1   *The conventional superhet architecture*

The superhet receiver has long been the default architecture of choice for the vast majority of applications. Moving the wanted signal to a fixed, intermediate frequency (IF) exchanges the difficult problem of making a tuneable channel filter for the generally rather easier one of making a tuneable local oscillator (LO). However, there is an issue with the so-called image signal which appears at a frequency on the opposite side of the LO and which needs to be removed with an RF filter before the mixer. When the wanted signal is at a high frequency and the channel bandwidth is narrow, it can be impossible to identify a single IF that allows a practical realisation of the RF and channel filters with the required degree of selectivity. Hence, as shown in Figure 2.1, there is often a need to use more than one IF to achieve the necessary receiver performance.

The superhet can offer excellent sensitivity and selectivity. However, its intrinsic reliance on highly selective, passive, analogue, off-chip filters usually makes it comparatively bulky and expensive to implement. It also tends to rely heavily on the use of automatic gain control (AGC). These are all serious impediments to multimode operation. There is also little scope for addressing this issue by increasing the level of digitisation as the analogue-to-digital conversion (ADC) would have to take place very late in the receiver chain, leaving most of the signal processing in the analogue domain.

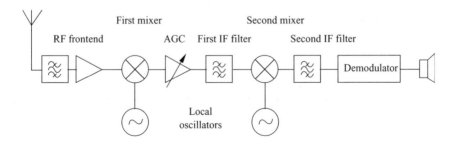

Figure 2.1   Conventional superhet architecture

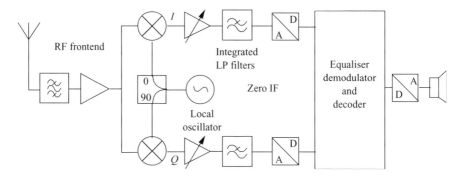

Figure 2.2   Zero-IF receiver

2.2.2   The zero-IF architecture

The zero-IF receiver [1] has obvious merits but it is only recently that it has been widely used for applications calling for good performance. By shifting the IF to zero, the need for any image filter is eliminated and the channel filter effectively becomes a pair of lowpass devices that can then be realised as active, on-chip circuits with minimum power consumption. There is the added complication of having to process the wanted signal as an *I* and *Q* complex pair to enable the resolution of positive and negative frequencies. However, on an integrated circuit this overhead is generally insignificant. It is also worth stressing that, in a zero-IF receiver, the LO is not phased locked to the wanted signal and, thus, even though the two receiver branches carry signals that are at baseband, there is still a demodulation process to be carried out. Figure 2.2 illustrates a zero-IF receiver implementation intended for GSM in which demodulation, together with equalisation and decoding, are all carried out in the digital domain. The fact that the channel filtering is done with integrated, low-pass filters already provides a degree of improved flexibility, even if these functions remain in the analogue domain.

The zero-IF receiver does suffer from some serious limitations. These are mainly associated with unwanted DC offsets and second-order intermodulation products generated in the mixers which occupy the same frequency band as the wanted signal. They can seriously desensitise the receiver and are particularly troublesome in narrowband systems. Nevertheless, some handset manufacturers and silicon suppliers have developed successful zero-IF products for GSM.

By contrast, the zero-IF receiver is ideally suited to wideband systems, such as UMTS. This is because the use of spread-spectrum techniques makes it easier to filter out the DC offsets and some of the second-order intermodulation products without having a significant impact on receiver sensitivity.

2.2.3   Low-IF architecture

At first glance, the low-IF receiver [2, 3], illustrated in Figure 2.3, looks similar to a zero-IF receiver. The crucial difference, however, is that the LO, rather than

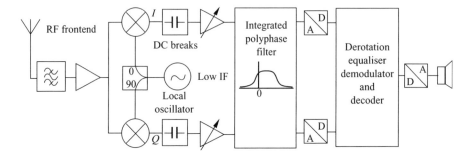

Figure 2.3   Low-IF architecture

being positioned at the centre of the wanted signal, is offset by half the channel spacing. In the context of GSM, this increases the IF from zero to 100 kHz. The main justification for this is to separate the wanted signal from the unwanted DC offsets and second-order intermodulation products. This allows the DC offsets to be managed by simple AC coupling without causing serious damage to the wanted signal. A significant proportion of the other second-order products is rejected by the channel filter. As will be described later, the effect can be to permit a substantial reduction in the second-order intercept point (IP2) requirement of the receiver. Alternatively, a low-IF receiver using the same mixers will be considerably more robust than a zero-IF receiver when operating in a hostile interference environment. This improvement comes about primarily as a result of the wanted signal having been separated from the second-order intermodulation products, rather than any filtering introduced by the DC breaks.

In an analogue implementation, the channel filtering is most conveniently realised by means of an active polyphase filter [4], as this provides the greatest immunity to the effect of random spreads in component values. Any consequent lack of image rejection due to imbalances in *I* and *Q* only affects one of the adjacent channels, where the selectivity requirements are generally not severe. Such a filter inevitably consumes more power than the two low-pass filters that would be required in a corresponding zero-IF receiver, but the overhead is generally modest. In the context of GSM, it is generally necessary to shift the IF back to zero, after the channel filter, and this derotation process, as shown in Figure 2.3, is best carried out in the digital domain.

2.3   Digitising a conventional low-IF architecture

The diagram given in Figure 2.4 illustrates a low-IF receiver architecture [5] with a still higher level of digitisation than that shown in Figure 2.3. At the frontend it comprises the usual combination of RF filter, low-noise amplifier and quadrature down-converter for capturing the incoming RF signals and placing the wanted channel on the IF of 100 kHz. To remove the troublesome DC offsets generated in the mixers,

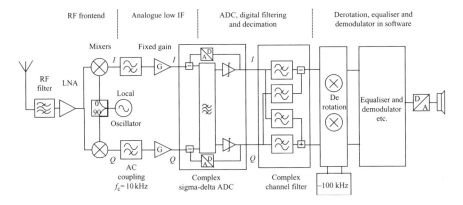

Figure 2.4   Complex signal processing in a low-IF receiver architecture

the I and Q components of the IF signal pass through a pair of AC couplings which act as first-order highpass filters with a cut-off frequency of approximately 10 kHz. The cut-off frequency is not critical but this value is high enough to allow the removal of DC offsets without damaging the wanted signal. As shown in the figure, the IF signals are amplified by two fixed-gain elements and then fed directly into a complex ADC. The need for AGC has therefore been eliminated. The ADC is complex to align the spectrum of its quantisation noise with that of the wanted signal. It is realised as a pair of cross-coupled $\Sigma\Delta$ modulators which digitise the I and Q inputs into a pair of bit streams of 1-bit resolution. These are in turn filtered by the digital channel filter whose dual roles are to remove any external interferers accompanying the wanted signal and also to remove the high-frequency quantisation noise generated by the ADC.

There can be little doubt that eliminating the need for analogue channel filtering and AGC in this way is a significant step forward in terms of giving the receiver better multimode potential. However, moving the ADC function to such an advanced position in the signal chain carries the disadvantage of increasing the dynamic range requirement of the ADC. As will be explained later, the dynamic range required for GSM in these circumstances is in the region of 100 dB and when this is combined with the complex aspect of the signal processing, the design of the ADC becomes a major challenge. There is also the disadvantage of needing a total of four FIR filter blocks in order to realise the asymmetrical frequency characteristics of the complex channel filter.

2.4   Performance requirements

The GSM standard is based on a mixture of time-division multiple access (TDMA) and frequency-division multiple access (FDMA). Each user is allocated a particular frequency channel of width 200 kHz and a time slot of duration 576.9 μs. Over the frequency range 925–960 MHz there are 175 available channels and over a period of

4.16 ms there are eight available time slots. The information rate is typically 9.6 kb/s which, after channel coding and training bits are added, translates into a physical bit rate of 270.833 kb/s. The type of modulation used is Gaussian minimum shift keying (GMSK) with a BT factor of 0.3. Responsibility for the GSM standard now lies with the 3rd-Generation Partnership Project (3GPP) [6].

Full details of the performance requirements relevant to the design of the receiver can be found in Reference 7. These relate to a 'small' mobile terminal receiving full-rate speech. There are several different fading conditions under which the terminal must operate, the most relevant of which are the static, TU50 and EQ50 profiles. With a static profile (i.e. no fading) the requirement on bit error rate (BER) is generally 2 per cent unless specified otherwise. For the TU50 profile, which is intended to represent a typical urban situation where the terminal is limited to a velocity of 50 km/h, the BER requirement is 8 per cent. A 3 per cent BER usually applies to the EQ50 profile. The most important of the receiver characteristics for GSM are listed in Table 2.1.

It should be noted that although the GSM specification only calls for a reference sensitivity of -102 dBm, most receivers in current use actually achieve a sensitivity of -108 dBm and therefore this is now the more relevant performance objective. The AM interferer test is intended to verify that a mobile terminal will continue to function in close proximity to a GSM basestation belonging to another operator. The bursting of the large signal from the basestation can cause second-harmonic distortion in the receiver frontend which is particularly damaging if the rising edge of the burst coincides with the training sequence in the middle of the burst of the wanted signal. Zero-IF receivers are particularly prone to this kind of interference since the intermodulation products are mostly located around DC. Low-IF receivers, on the other hand, are more resilient because of the offset of the wanted signal, as previously explained.

Table 2.1   Receiver characteristics

Maximum input	-15 dBm for 0.1 per cent BER, static channel
Reference sensitivity	-102 dBm for 8 per cent BER, TU50 channel
	(c.f. state of the art -108 dBm)
Cochannel interferer	-9 dB with respect to wanted signal at -82 dBm,
	8 per cent BER, TU50 channel
Adjacent-channel interferer	9 dB with respect to wanted signal at -82 dBm,
	8 per cent BER, TU50 channel
Alternate-channel interferer	41 dB with respect to wanted signal at -82 dBm,
	8 per cent BER, TU50 channel
Largest blocking signal (CW)	-23 dBm, 3 MHz offset, wanted signal at -99 dBm,
	2 per cent BER, static channel
AM interferer	-31 dBm, 6 MHz offset, wanted signal at -99 dBm,
	2 per cent BER, static channel

2.5   Receiver dimensions

The receiver noise figure derives largely from the system requirement for sensitivity. The logical starting point for the calculation is the estimation of the signal-to-noise ratio (SNR) needed at the demodulator/equaliser of a pseudo-ideal receiver. According to basic system simulations, the E_b/N_0 required for an 8 per cent BER with a TU50 propagation channel is 5.2 dB. When a correction factor of 1.32 dB is applied to take account of the ratio of the bit rate to the channel bandwidth (i.e. 270.833/200), the SNR estimate becomes 6.5 dB. Hence if the desired sensitivity, P_{sen}, is -108 dBm, the total noise power at the input to the receiver under these conditions, P_n, must be given by:

$$P_n = P_{sen} - SNR_{gmsk} = -108 - 6.5 = -114.5 \text{ dBm.} \tag{2.1}$$

If this is compared with the thermal noise level (-173.83 dBm) raised by 53 dB to account for the 200 kHz channel bandwidth, the overall receiver noise figure, F_{dB}, must then be:

$$F_{dB} = P_n - 10\log(200 \times 10^3) + 173.83 = 6.5 \text{ dB.} \tag{2.2}$$

The frontend alone needs a noise figure slightly less than this to allow for the quantisation noise contribution of the ADC. By assigning the frontend a noise figure of 6.0 dB, the degradation in receiver noise figure caused by the ADC must be approximately 0.5 dB, which corresponds to a referred ADC noise power, P_{adc}, of:

$$P_{adc} = P_n - 0.5 + 10\log_{10}[10^{0.5/10} - 1]$$
$$= -114.5 - 0.5 - 9.136 = -124.1 \text{ dBm.} \tag{2.3}$$

In further references to this parameter the value will be rounded down to -125 dBm. A graphical illustration of the preceding calculations is given in Figure 2.5. One final step is required if proper account is to be taken of the signal losses in the RF frontend. With an estimated 2.5 dB losses in the passive RF parts including the band selection filters and antenna feed arrangement, the noise figure of the main active part of the receiver must be further reduced to a value of 3.5 dB.

The adjacent-channel and alternate-channel rejection requirements of the receiver are illustrated by the curves plotted in Figure 2.6. These represent the spectral envelopes of a wanted signal, a cochannel, an adjacent-channel and an alternate-channel interferer at the relative frequencies and levels stipulated in the GSM specification. For the tests in question, the wanted signal is at a relatively high level of -82 dBm, at which point the receiver noise can be ignored. Given that the receiver can achieve an 8 per cent BER in a TU50 propagation channel with an SNR of 6.5 dB, there is generally no difficulty in meeting the cochannel rejection requirement of -9 dB. Then, on the basis that an adjacent-channel interferer at a level $+9$ dB above the wanted signal must be attenuated to the level of a cochannel interferer, the adjacent-channel rejection requirement must be at least 18 dB. Similarly, the alternate-channel rejection requirement must be in the order of 50 dB.

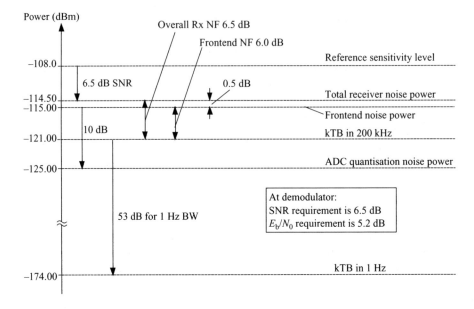

Figure 2.5   Calculating noise figure from reference sensitivity

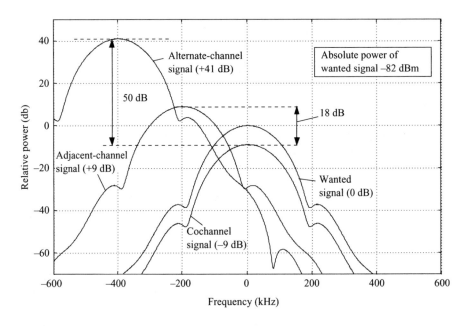

Figure 2.6   Illustrating channel filter rejection requirements

Of the different blocking signals referenced in the 3GPP specification, it is the signal at a 3 MHz offset with a level of −23 dBm (P_{int}) that is generally regarded as the most demanding. Under the relevant test conditions the wanted signal is at a level of −99 dBm and if the required SNR of 6.5 dB is to be achieved, the residual power of the blocking signal after filtering must be no higher than −(99+6.5) = −105.5 dBm. At the level of −114.5 dBm, the receiver noise power can largely be ignored and the rejection requirement for the blocking interferer, A_{blk}, is then simply:

$$A_{blk} = P_{int} - P_{sig} + SNR_{gsm} = -23 + 99 + 6.5 = 82.5 \text{ dB}. \tag{2.4}$$

The dynamic range requirement of the ADC in the receiver is determined by the level of this same blocking interferer and by the permitted level of the ADC quantisation noise, P_{adc}. Hence, the difference between the −23 dBm level of the blocking signal and the −125 dBm level of the noise gives a dynamic range requirement of 102 dB. Although substantial, this is known to be within the capability of the $\Sigma\Delta$ modulator type of ADC to be described. The design problem is also not quite so severe if the SINAD (SIgnal-to-Noise-And-Distortion) requirement of the ADC is taken into consideration which, for the purposes of passing the blocking-interferer test, is a lower value of 85.5 dB. This is derived in much the same way as the filtering requirement for the blocking signal, i.e.

$$\text{SINAD} = P_{int} - P_{sig} + SNR_{gsm} + 3 = -23 + 99 + 6.5 + 3 = 85.5 \text{ dB}, \tag{2.5}$$

the extra 3 dB being inserted to allow both the ADC noise and the residue of the blocking signal to contribute equally towards the maximum permitted noise residue mentioned above of −105.5 dBm.

The AM interferer test can be translated into an IP2 requirement using a mixture of simple reasoning and system simulation. If the wanted signal is at a level of −99 dBm and the SNR requirement for a 2 per cent BER under static fading conditions is 6.5 dB, it follows that the maximum tolerable power level of the second-order intermodulation products (referred back to the receiver input) is −105.5 dBm. With an interferer power level of −31 dBm, some simple linear extrapolation yields an initial estimate for the IP2 requirement of a zero-IF receiver of +43.5 dBm. It remains then to estimate how much this is reduced for a low-IF receiver by virtue of the filtering effects of the AC coupling and channel filter. The most reliable way to estimate this reduction is to perform a system simulation for a receiver whose IF can be changed and whose second-order non-linearity can be adjusted. For the sake of brevity, the details of the simulation will not be described here but the results are summarised in Figure 2.7. The IP2 value obtained for the zero-IF receiver was 42.5 dBm, which is close to the original, crude estimate of 43.5 dBm. This corresponds to a maximum tolerable level of second-order intermodulation distortion of −104.5 dBm, suggesting that the receiver is just meeting specification at an SNR of 5.5 dB. More interestingly, the simulation suggests that the low-IF receiver can meet the AM interferer test with an IP2 of only +28.5 dBm, which is a very worthwhile improvement of 14 dB on the zero-IF receiver. This could be used to reduce power consumption or give the receiver much greater immunity to the AM interference.

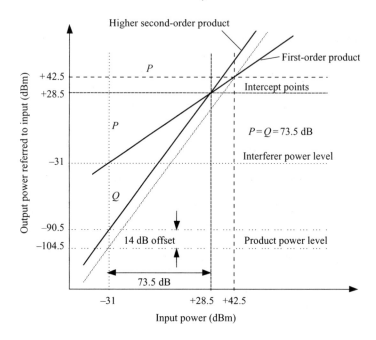

Figure 2.7   IP2 requirements for zero- and low-IF options

2.6   The change to non-complex signal processing

A diagram of the revised low-IF receiver architecture that avoids the need for a complex ADC is given in Figure 2.8. The basic idea is that instead of processing both the I and Q components of the low-IF output from the frontend, the Q component is discarded and only the I component is passed to a single, non-complex, $\Sigma\Delta$ modulator. The effect of this is to make the spectrum of the now real IF signal symmetrical about zero, as if the IF were itself zero (see the spectrum at B in Figure 2.8). Hence, the spectrum is naturally aligned with that of the quantisation noise spectrum of the single $\Sigma\Delta$ modulator. It is then relatively easy to make the IF signal complex once again by performing the equivalent of a Hilbert transform in the digital channel filters that follow. The rest of the signal chain is the same as the previous system of Figure 2.4.

An obvious consequence of processing only the I component of the IF signal is a complete loss of image rejection. Any interferer present in the lower adjacent channel will be folded over about zero frequency and become indistinguishable from the wanted signal. There is also a problem with the lower alternate channel as when this signal becomes folded about zero it falls into what is effectively the adjacent channel on the upper side of the wanted signal. In this position the lower tail of its spectrum falls into the band of the wanted signal and due to its relatively high power level the interference caused can degrade receiver sensitivity. Fortunately, both these problems can be easily overcome by filtering the output of the frontend with a simple polyphase filter. The filter attenuates any interferers present in the lower adjacent and

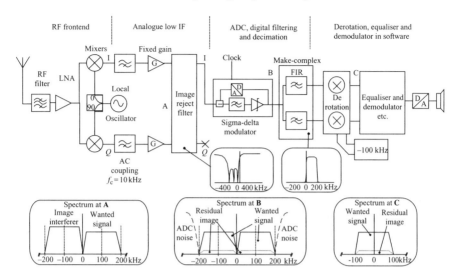

Figure 2.8   Non-complex signal processing in a low-IF receiver architecture

alternate channels to a sufficiently low level that they no longer have any significant effect. Except for some interstage buffers, the polyphase filter is a passive device and therefore consumes little DC power.

Like the original system of Figure 2.4, the system of Figure 2.8 uses no AGC and despite the presence of the polyphase filter the vast majority of the channel selectivity is implemented in the digital domain. The DC offsets from the mixers are removed by AC couplings, whose highpass characteristics have the same 10 kHz cut-off frequency as used previously. The single $\Sigma\Delta$ modulator addresses the same dynamic range as before but is now very much easier to design for optimum conversion efficiency and for improved adaptability. As will be shown, the dynamic range can be achieved with a fifth-order $\Sigma\Delta$ modulator whose clock frequency is 26 MHz. The high levels of quantisation noise produced by the modulator outside the bandwidth of the wanted signal are heavily attenuated by the subsequent digital filters. This pair of FIR filters also provides the necessary channel filtering as well as reconstructing the Q component to make the IF signal complex once again. Only two FIR blocks are required because of the absence of a Q input. It is also important to note that the blocks decimate by a factor of 96 in one step, lowering the sampling frequency from 26 MHz at the input, to the bit rate of 270.8333 kHz at the output. Decimating in one step improves power efficiency by avoiding the need for digital multiplication. The signal processing then involves only changing the sign of the tap weights prior to summation, in response to changes in the state of the binary input bit stream from the $\Sigma\Delta$ modulator.

Whilst improved adaptability is the main objective behind this non-complex signal processing, there will usually be some advantage to be gained in terms of power consumption over the alternative of a fully-complex architecture. To explain this it is helpful to split the power consumption of the whole ADC function into two parts, namely that associated with only the $\Sigma\Delta$ modulators, P_1, and that associated with the decimation filters, P_2. Furthermore, to establish a point of reference in this elaboration, the two powers P_1 and P_2 will be defined as the powers consumed by the two parts of the ADC function in an equivalent zero-IF receiver consuming a total power P_{zif}. In this scenario it is common for P_1 and P_2 to be of approximately the same magnitude.

To make the step from a zero-IF to a conventional low-IF configuration, some extra cross-branch circuitry is needed between the loop filters of the two $\Sigma\Delta$ modulators to make the pair complex. There will be a very slight increase in power consumption associated with this but the increase is so small as to be ignored. Given the already large over-sampling ratio, there should be no need to change either the clock speed or the bandwidth of the basic loop filters and hence the change in power consumption in switching from a zero-IF to a low-IF for the pair of $\Sigma\Delta$ modulators should be insignificant. As far as the digital filters are concerned, however, moving to the low-IF configuration will double the number of FIR filters required to provide the complex filtering function and hence for the same clock speed, the power consumption will typically double. This means that the power consumption for the conventional low-IF receiver, P_{nzif1}, will be approximately 50 per cent greater than for the direct-conversion receiver, i.e.

$$P_{nzif1} = P_1 + 2 \times P_2 = P_{zif} \times 150 \text{ per cent.} \tag{2.6}$$

In moving to the low-IF receiver with non-complex signal processing, there is a need to double both the bandwidth of the loop filter in the $\Sigma\Delta$ modulator and the clock speed in order to maintain the same dynamic range in the band 0–200 kHz. In a practical $\Sigma\Delta$ modulator, the power consumption and the noise output in the band of the wanted signal are dominated by the dimensions of the input stages of the loop filter and have only a weak dependency on the clock speed. If the loop filter is modified by just halving the area of the capacitors and not by any change in the size of the active devices, there should be virtually no change in its power consumption. Hence, in needing only one instead of two modulators, there should be a saving of 50 per cent in the power consumption figure P_1. Unfortunately, the same is not true of the power consumption of the digital filters since despite the need for only two FIR filter blocks, doubling the clock speed doubles the number of taps required. This means the value of P_2 remains unchanged. If this analysis is valid, it can be concluded, therefore, that the power consumption of the ADC function in the non-complex, low-IF receiver (P_{nzif2}) will be at least 25 per cent higher than that for the direct-conversion receiver but it could be of the order of 17 per cent lower than that of the fully complex low-IF receiver, i.e.

$$P_{nzif2} = P_1/2 + 2 \times P_2 = P_{zif} \times 125 = P_{nzif1} \times 83 \text{ per cent.} \tag{2.7}$$

This power saving should be worthwhile but remains to be fully verified in practice.

2.6.1 Polyphase image-rejection filter

Passive polyphase filters were studied some time ago by Gingell [8] and as a prototype structure take the form shown in Figure 2.9. They are a cascade of RC sections each one of which is capable of creating a transmission zero at a frequency, ω_{zi}, such that $\omega_{zi} = -1/R_iC_i$. Whether the zero is at a positive or negative frequency will depend upon the relative polarities of the I and Q components of the input voltage. If each section were to be treated in isolation and driven by a pair of voltage sources, the frequency of the transmission zero would uniquely define the whole of the frequency response and the impedance of the section would be of no consequence. However, because of the loading effect of each successive RC section on its preceding neighbour, the overall frequency response does become affected by the relative impedances and this substantially complicates the synthesis process. As a general rule of thumb, the impedance of the sections should increase in the direction of the output to minimise the loading effects but it is usually necessary to adjust the impedances in a process of trial and error to achieve the most desirable frequency response. Hence, the design procedure involves choosing the transmission zeros to give the desired stopband response and the impedance levels to give the desired passband response. The desired passband response is one which is as flat as possible at positive frequencies close to zero in the region of the wanted signal.

Without the polyphase filter, the modified receiver would have no rejection in the lower adjacent channel and therefore the polyphase filter must provide all of the 18 dB required over the band -200 kHz to zero. If the problem with the lower alternate channel is to be avoided it must also provide at least 32 dB of attenuation in

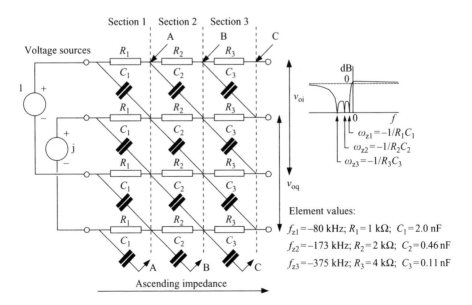

Figure 2.9   Passive polyphase filter prototype network

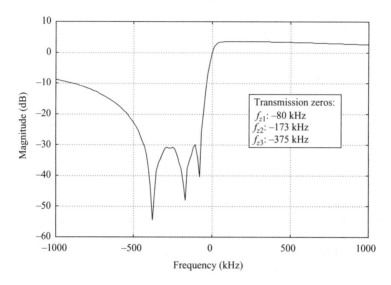

Transmission zeros:
f_{z1}: −80 kHz
f_{z2}: −173 kHz
f_{z3}: −375 kHz

Figure 2.10   Frequency response of passive polyphase filter

the band −400 kHz to −200 kHz. This brings any interferer in the alternate channel at a relative level of +41 dB with respect to the wanted signal down to the maximum permitted level of an adjacent-channel interferer at +9 dB (Figure 2.6). The remainder of the receiver selectivity can then be provided by the digital channel filters. For the purposes of designing the polyphase filter, the attenuation requirement is assumed to be at least 32 dB across both the lower adjacent and alternate channels as this carries no practical penalties and gives some useful performance margin.

To obtain the required attenuation over the approximate band −400 kHz to zero, the transmission zeros for the prototype network of Figure 2.9 are set at the frequencies −80 kHz, −173 kHz and −375 kHz whilst the corresponding resistor values are 1 kΩ, 2 kΩ and 4 kΩ. On analysis with a circuit simulator, the network produces a frequency response of the form plotted in Figure 2.10 in which the stopband requirement is largely fulfilled over the band −400 kHz to −60 kHz. The response in the passband at positive frequencies between about 30 kHz and 1 MHz is also flat as required, the slight positive gain (4 dB) being a consequence of the impedance transformations in the filter and the use of ideal voltage sources. Difficulties with physical implementation and with group delay variation prevent the transmission zero at −80 kHz being moved closer to zero but the lack of rejection in this region has no significant impact on the ability of the network to reject an adjacent-channel interferer. In any case, the AC coupling elements elsewhere in the receiver chain will force a transmission zero at DC.

It should be noted that in a practical implementation, the polyphase filter incorporates interstage buffer amplification to help reduce the loading effects of successive stages and to render the thermal noise generated by its resistors insignificant. The power consumption of the buffers is in the region of 10 mW which is a very small fraction of the total power consumption of the RF frontend (typically 135 mW). In the forthcoming system simulations, the filter is modelled as a complex FIR device whose impulse responses are 500 samples in length at a sampling frequency of 26 MHz. This represents a significant computational overhead but is of no relevance to the practical implementation of the filter.

2.6.2 Make-complex channel filter

As previously stated, the output of the single $\Sigma\Delta$ modulator must be made complex once again and this is accomplished with a pair of FIR filters that perform the equivalent of a Hilbert transform [9]. A Hilbert transform of a real function in time $I(t)$ is given by:

$$Q(t) = \frac{1}{\pi} \int_{-\infty}^{\infty} \frac{I(\tau)}{t - \tau} d\tau \qquad (2.8)$$

where $Q(t)$ has a frequency spectrum whose amplitude components are the same as $I(t)$ but whose phase components are shifted by $-90°$ for positive frequencies and $+90°$ for negative frequencies. In the time domain, it can be interpreted as a convolution of $I(t)$ with the function $1/(\pi t)$ whose Fourier transform has a flat amplitude response over all frequencies but a phase response which has a step discontinuity between $+90°$ and $-90°$ at zero frequency. Hence the transform implements an ideal highpass filter whose cut-off frequency is infinitesimally close to zero.

In practice, a close approximation to the transform can be implemented with a pair of filters of the form illustrated in Figure 2.11. Here the FIR filter in the I signal path has a lowpass amplitude response, offering the additional benefit of some high-frequency selectivity. The FIR filter in the Q signal path has an identical amplitude

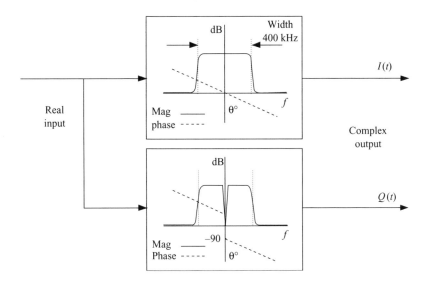

Figure 2.11   Digital make-complex and channel filter arrangement

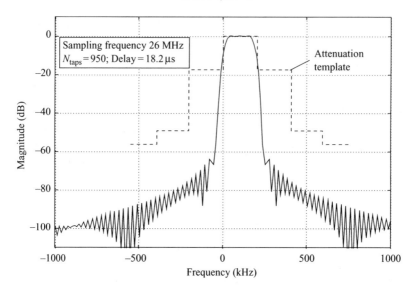

Figure 2.12   Frequency response of digital filter

response except for the presence of a hole at DC. The width of the hole is determined by the sampling rate and by the length of the impulse responses of the filters. The phase characteristic for the *I* component is preferably linear and has a slope determined by the delay of the filter. The phase characteristic of the *Q* component is identical except for the step discontinuity in the middle. Both filters must have the same delay.

Whilst a suitable complex filter function can be derived by separate treatment of the two FIR components, the frequency response plotted in Figure 2.12 was obtained by a slightly more direct route whereby a real, lowpass response was first synthesised with an appropriate shape and then simply translated in frequency by +100 kHz. It is then possible to calculate the two respective impulse responses for the *I* and *Q* components (Figure 2.13) using an inverse, discrete Fourier transform. The delay of the filter is 18.2 μs, which is equivalent to 475 samples at a sampling frequency of 26 MHz. As shown in Figure 2.12, the filter is substantially more selective than the attenuation template would suggest is necessary but the extra selectivity and deep stopband floor are essential to attenuate the high levels of quantisation noise generated by the $\Sigma\Delta$ modulator.

2.6.3 $\Sigma\Delta$ modulator ADC

$\Sigma\Delta$ modulators and their use as ADCs are discussed in some detail in References 10 and 11. As explained, they offer high conversion efficiency, excellent linearity and strong immunity to aliasing due to their high over-sampling ratio. Their basic principle of operation is illustrated by the diagram given in Figure 2.14. There are three main parts to the modulator: the loop filter in the forward signal path; the output limiter; and the digital-to-analogue converter (DAC) in the feedback path. Signals

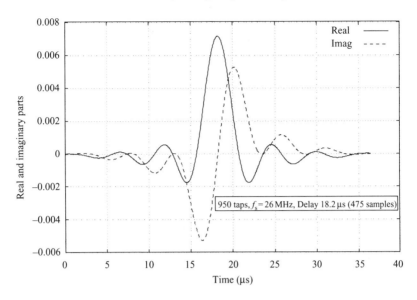

Figure 2.13   Impulse responses of digital filter

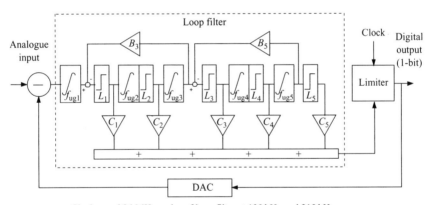

Clock speed 26 MHz, poles of loop filter at 130 kHz and 210 kHz

Figure 2.14   Behavioural model of fifth-order sigma-delta modulator

entering the modulator are amplified by the loop filter and then quantised into a set of only two output levels (i.e. $+1$ and -1). The effect of the quantisation is to generate very large quantities of noise. However, by feeding the noisy output signal back into the input via the very simple DAC, the loop filter is able to alter the shape of the frequency spectrum of the noise and move most of its power to a very high frequency away from the vicinity of the wanted signal. It is then the task of the digital filter that follows to remove the quantisation noise and construct a multibit representation of the wanted signal with the required resolution.

The noise density at any point in the modulator output spectrum is a function of the sampling frequency and the gain of the loop filter. Hence, a high sampling frequency will reduce the noise density by spreading the noise over a wide bandwidth, whilst the high gain of the loop filter at low frequencies will give more effective cancellation and reduce the noise density in the region of the wanted signal. Typically, the loop filter comprises a cascade of integrators with unity-gain frequencies, f_{ugi}. If these are realised as time-continuous circuits the need for any anti-alias filters before the modulator can be largely avoided. In the fifth-order example shown in Figure 2.14, the loop filter is of the so-called 'feedforward' type in which the outputs from all five integrators are suitably weighted by coefficients, C_i, before being added together at the common output. Two pairs of integrators are surrounded by feedback elements, B_3 and B_5 which generate conjugate pairs of transmission poles at $\pm130\,\text{kHz}$ and $\pm210\,\text{kHz}$. The remaining integrator without feedback produces a single transmission pole at DC. Offsetting the poles in this way gives the loop filter a greater average gain over a 200 kHz bandwidth, leading to more effective noise shaping and a correspondingly lower level of in-band quantisation noise. Each integrator output passes through a clipping circuit that limits the output swing to a level of L_i. Their purpose is to help maintain loop stability under large-signal drive conditions by progressively reducing the effective loop gain. They also prevent the loop from entering a latch-up condition.

With suitable values for the set of parameters f_{ugi}, C_i, B_3, B_5 and the DAC gain, the fifth-order modulator produces an output spectrum of the form shown in Figure 2.15 when its input is a pure tone of 100 kHz. In this case the tone is at the maximum permissible drive level for the modulator, which corresponds to a relative amplitude of 0.7 or a relative power level of -6 dBm. Any higher than

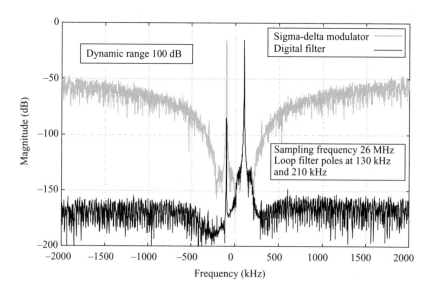

Figure 2.15   Output spectra from fifth-order sigma-delta modulator and digital filters

this and the modulator becomes unstable. The clock speed is 26 MHz. As shown in the figure, the tone is clearly present on both sides of the spectrum, as is to be expected with only the I component of the IF signal represented. The spectrum of the quantisation noise is also symmetrical about zero, the hole in the centre having a width of approximately 400 kHz. Figure 2.15 also shows the spectrum of the modulator output after it has passed through the pair of digital channel filters. Not only has the pair of filters eliminated the majority of the quantisation noise but it has also made the signal complex once more, as evidenced by the different levels of the tones now seen on either side of the spectrum. The tone on the right at $+100$ kHz is unchanged in magnitude but that on the left at -100 kHz has been attenuated by approximately 70 dB. This is the effective image rejection ratio of the make-complex filter function. By selective integration of the power in the spectrum, the dynamic range of the modulator and filter combination is found to be in the region of 100 dB.

2.7 System simulations

A series of system simulations has been carried out to verify the basic functionality of the modified receiver and its ability to meet the various GSM specifications. For increased confidence in the results, these simulations were carried out in two parallel exercises, one involving the coding of the various signal processing functions in FORTRAN and the other involving the use of the commercially available software SPW [12]. The relevant system block diagram for the simulations is given in Figure 2.16, the upper part of which is concerned with generating the receiver input signal, the middle part comprises the majority of the IF signal chain and the lower part deals with demodulation and equalisation. All the BER calculations are based on a comparison of physical (i.e. not coded) bits at the bit rate of 270.833 kb/s. As shown, the receiver input comprises a wanted signal centred initially on a zero IF and an interferer shifted by an arbitrary frequency offset, both of which pass through independent, fading channels. On entry into the receiver, Gaussian noise is added at a level of -115 dBm to take account of the 6.0 dB noise figure of the frontend. The whole spectrum is then translated by 100 kHz to position the wanted signal on the chosen IF.

All the filters of the receiver system are modelled as FIR devices, including the AC couplings whose cut-off frequency is 10 kHz. The gain of the amplifier shown at the input to the polyphase filter has a fixed value of 16 dB, which when the 4 dB gain of the polyphase filter and the 3 dB loss associated with dropping the Q channel are added, maps the maximum input level of a blocking signal at -23 dBm on to the -6 dBm maximum drive level of the single $\Sigma\Delta$ modulator.

The simulations for receiver sensitivity with a TU50 propagation channel yield the results plotted in Figure 2.17. As shown, without the $\Sigma\Delta$ modulator present, the BER falls through a value of 8 per cent for an input signal power of -108.5 dBm and continues to fall towards zero at an input power of -82 dBm. It remains zero for all higher input powers since without the $\Sigma\Delta$ modulator the receiver is perfectly linear. With the $\Sigma\Delta$ modulator present, the degradation in receiver sensitivity for a BER of 8 per cent is less than 0.5 dB, confirming that the quantisation noise power generated

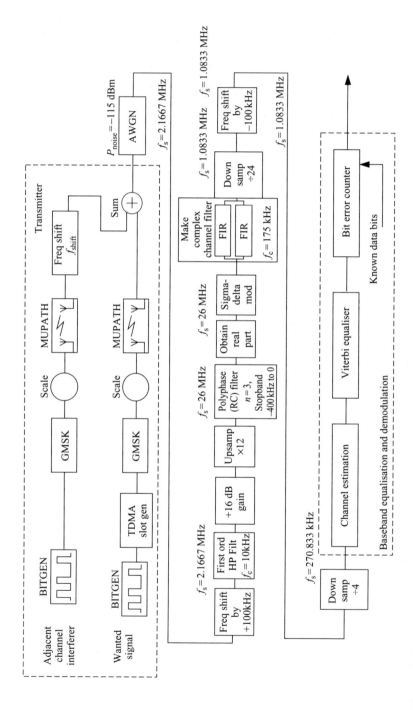

Figure 2.16   Block diagram for simulations

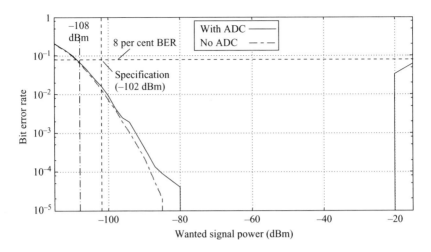

Figure 2.17   System simulations for receiver sensitivity

by the modulator must be at or below the intended value of -125 dBm. Hence, in terms of sensitivity the receiver achieves its target performance of -108 dBm. At higher signal levels the $\Sigma\Delta$ modulator appears to cause some minor degradation in BER for input powers in the region of -80 dBm. This effect is believed to be linked to the non-linear behaviour of the channel equaliser for very low BERs, rather than just a simple loss of signal-to-noise ratio. In any case it is of little consequence. At still higher power levels the BER is substantially zero until it rises sharply at an input power of -20 dBm. The rise in BER is caused by an overdrive of the $\Sigma\Delta$ modulator. As the receiver must operate with a maximum input signal level of -15 dBm (static channel) this rise in BER is slightly premature. However, there is some doubt over the accuracy of the behavioural model used for the $\Sigma\Delta$ modulator when operating under such large-signal conditions and results obtained in practice suggest that a real receiver is more likely to meet this particular performance requirement. If it does not then it would not be difficult to increase the dynamic range at the high end by introducing a simple one-step change in the gain of the frontend.

Figure 2.18 illustrates the selectivity performance of the receiver in dealing with cochannel, adjacent-channel and alternate-channel interferers. In each case, the BER for a TU50 propagation channel is plotted against the relative level of the interferer whilst the wanted signal is held at a constant level of -82 dBm. For the cochannel interferer, the receiver passes the GSM specification by a margin of 1.5 dB, thereby confirming that the passband of the channel filters is well chosen and not unduly narrow. In the presence of the adjacent-channel interferers the receiver passes the specification by a margin of approximately 6 dB, the rejection of the interferer on the lower side being provided largely by the action of the polyphase filter and that on the upper side largely by the digital channel filters. For the alternate-channel interferers, the margin on the specification is even greater at a value of about 14 dB.

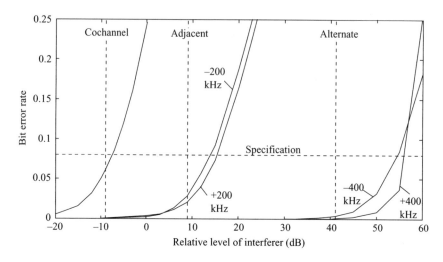

Figure 2.18   System simulations for selectivity

This is despite the overdrive of the $\Sigma\Delta$ modulator that is the principal cause of the rise in BER at a relative interferer level of $+55$ dB.

2.8   Incorporation in a dual-mode receiver

The low-IF architecture as described with its non-complex signal processing is a practical proposition for use with GSM and other such narrowband, FDMA/TDMA radio standards where a spread-spectrum air interface is not a feature. Depending upon the physical implementation, it could offer a modest saving in power consumption of the order of 17 per cent. However, by far the most important factor in its favour is the added flexibility that it can give to a highly digitised, multimode receiver that must operate with both spread-spectrum and non-spread-spectrum RF signals.

An illustration of how the modified, low-IF architecture can be incorporated in a dual-mode receiver is given in Figure 2.19. This particular receiver [13] is intended to operate in UMTS and GSM modes but is adaptable enough to accommodate other modes like CDMA2000 should this be necessary. It should be noted that there is a difficult design problem to be addressed in making the RF frontend cope with the multiple frequency bands that are encountered with the different modes. This is something of a separate problem from the design of the IF and baseband parts of the receiver but nonetheless is one that must be solved. For the rest of the receiver, however, the most important design issues relate to cost, complexity and power consumption. These translate into the need for an architecture comprising a minimum number of common hardware and software blocks which can be both **re-used** and **reconfigured** for use in all of the different modes. As previously explained, reconfiguration is generally easier if the majority of the blocks are digital in nature (see the forward location of the ADC(s) in the diagram). Re-use and reconfiguration were central principles

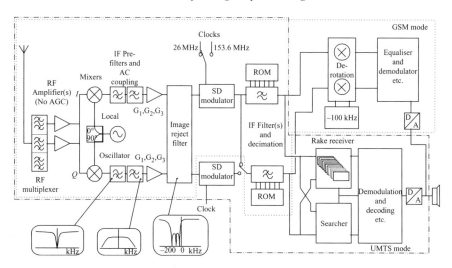

Figure 2.19   Dual-mode UMTS/GSM receiver architecture

behind the creation of the architecture of Figure 2.19 and the incorporation of the
non-complex IF processing.

With reference to Figure 2.19, the two dotted outlines indicate the respective
grouping of blocks used for the UMTS and GSM modes. Most of the blocks between
the RF frontend and the baseband signal processing parts have been re-used and even
some of the baseband parts, shown separate in the figure, could be implemented
on a common processor core. Depending upon the multiband capability of the RF
frontend, one or more quadrature pairs of mixers would translate the incoming signals
down to a zero-IF for UMTS and a low-IF of 100 kHz for GSM. The subsequent AC
couplings which have a cut-off frequency of 10 kHz to suit GSM, can stay in place
for UMTS without causing any detrimental effect. Similarly, the lowpass pre-filters
which have a cut-off frequency of 8.5 MHz to suit UMTS, can stay in place for GSM.
The IF pre-amplifiers shown have three gain settings constituting a small amount of
AGC. Two of the steps are required to cope with the full extent of the dynamic range
of the UMTS signals and are selected on the basis of the power detected at the output
of the digital channel filters. The third gain setting simply ensures that the different
power range encountered with a GSM signal is aligned with the signal-processing
window of the single $\Sigma\Delta$ modulator. In the case of the image rejection filter, this is
only required for the GSM mode and is preferably bypassed for UMTS. However, as
implied by the figure, it may be possible to leave it in place for UMTS with relatively
little impact on sensitivity.

Both $\Sigma\Delta$ modulators are active in the UMTS mode and clocked at a frequency
in the region of 150 MHz. This high clock frequency is necessary to ensure the
modulators cope with the increased bandwidth of a UMTS signal (i.e. in the region
of 4 MHz). They are fifth-order devices with a two-level output of the type already
described in Figure 2.14, the loop filter having been scaled to provide a different

pair of pole frequencies. Each one feeds its own digital, FIR filter which for UMTS has a root-raised-cosine (RRC) frequency response. When operating in the GSM mode, only one of the modulators is active, the other being shut down to save power consumption. In switching from UMTS mode, the working modulator is electronically reconfigured to change its loop filter bandwidth, whilst the clock frequency is changed to 26 MHz. Its output is then fed to both digital filters, whose frequency responses are also changed to implement the make-complex and channel filtering requirements for GSM. Changing the channel filter responses is a simple matter of re-addressing the appropriate set of tap coefficients stored in local memory.

In summary, the non-complex signal processing of the sort described in this chapter can give added flexibility to a multimode receiver. In particular, it allows a single, real $\Sigma\Delta$ modulator to be used as an ADC in place of what would otherwise be a fully complex modulator in a conventional low-IF receiver. The benefit has been demonstrated in the receiver structure shown in Figure 2.19, wherein operation in UMTS mode favours a zero IF and operation in GSM mode a low IF. Not only does this allow the same basic $\Sigma\Delta$ modulator to be used for both modes but it substantially simplifies the design of the modulator, already made difficult by the performance requirements of UMTS.

2.9   Conclusions

This chapter has presented a highly digitised, low-IF receiver architecture intended for use in a cellular radio application such as GSM. Its novel feature is that by processing only the real part of the complex IF signal after the mixer, it avoids the need for a complex ADC, thereby resulting in a very much simpler and more adaptable ADC circuit design. Advancing the position of the ADC in a low-IF receiver already gives improved adaptability by moving the channel filtering into the digital domain. This overall adaptability is further improved in the proposed architecture, as well as offering the opportunity for a modest saving in power consumption. The non-complex signal processing causes an inevitable loss in image rejection but this problem is easily overcome by including a simple passive polyphase filter before the ADC. The ADC itself is realized as a fifth-order $\Sigma\Delta$ modulator running with a clock speed of 26 MHz. Its bit-stream output is filtered extremely efficiently by a pair of digital FIR filters whose task is also to reconstruct the Q component of the complex baseband signal. System simulations demonstrate the effectiveness of the overall signal processing strategy and confirm that the level of performance needed to realize a competitive GSM product can be readily achieved. The application of the new architecture in a multimode receiver for GSM and UMTS has also been discussed.

2.10   References

1 RAZAVI, B.: 'Design considerations for direct-conversion receivers', *IEEE Transactions on Circuits and Systems-II*, 1997, **44** (6), pp. 428–35

2 MINNIS, B. J., MOORE, P. A., PAYNE, A. W., CASWELL, A. C. and BARNARD, M. E.: 'A low-IF polyphase receiver for GSM using log-domain signal processing'. IEEE RFIC2000 Symposium, Boston, MA, June 11–13, 2000, pp. 83–6

3 DROINET, Y.: 'Advanced RF technologies for the wireless market', *Microwave Journal*, September 2001, pp. 148–59

4 VOORMAN, J. O.: 'The gyrator as a monolithic circuit in electronic systems'. PhD Thesis, Catholic University of Nijmegen, The Netherlands, June 1977

5 ALI, D.: 'Radio Receiver'. International patent application No. WO 00/22735, September 1999

6 http://www.3gpp.org

7 ETSI Secretariat, 'GSM: digital cellular telecommunications system (phase 2) radio transmission and receptions (GSM 05.05)'. ETS 300 577, F-06921 Sophia Antipolis, Cedex, France, March 1996

8 GINGELL, M. J.: 'Single sideband modulation using sequence asymmetric polyphase networks', *Electrical Communications*, 1973, **48** (1 & 2), pp. 21–5

9 HAYKIN, S.: 'Communication systems' (Second Edition, John Wiley, 1983)

10 BOSER, B. E. and WOOLEY, B. A.: 'Design of sigma-delta modulation analogue-to-digital converters', *IEEE Journal of Solid-State Circuits*, 1988, **23** (6), pp. 1298–308

11 CANDY, J. C. and TEMES, G. C.: 'Oversampling methods for A/D and D/A conversion', 'Oversampling Delta-Sigma Data Converters, Theory, Design and Simulation' (IEEE Press, Piscataway, N.J., 1991) pp. 1–25

12 Cadence Design Systems, 'System processing worksystem (SPW)', http://www.cadence.com

13 MINNIS, B. J. and MOORE, P. A.: 'A reconfigurable receiver architecture for 3G mobiles'. 2002 IEEE Radio Frequency Integrated Circuits (RFIC) Symposium, Seattle, WA, June 2–4 2002, pp. 187–90

Chapter 3

A reconfigurable baseband chain for 3G wireless receivers

Md. Iqbal Younus, Anup Savla, Arun Ravindran and Mohammed Ismail

3.1 Introduction

The rapid growth of wireless communications has led to the proliferation of different standards. The highly competitive market demands low cost, low power and small form-factor devices. Convergence of voice, data and video, evolving standards and a high demand for new features as they are invented, all require equipment vendors to build systems that are flexible and field upgradeable [1]. In such an environment where requirements are rapidly changing, the time it takes to bring new products to market can spell the difference between success and failure. Vendors need an approach that allows them to quickly design, debug and verify their systems. This chapter discusses modular 3G baseband building blocks with a focus on the design of a reconfigurable baseband chain that precedes the analogue-to-digital converter (ADC) in a multistandard fully integrated wireless receiver. The circuits proposed here retain flexibility and simplicity of design, and all are based on one active element. Using these circuits, a digitally reconfigurable baseband chain is implemented. The baseband chain is adapted to accommodate GSM, IS-95, and WCDMA wireless standards. The proposed approach enables the development of highly programmable, flexible and upgradeable baseband circuits with a short design cycle.

3.2 Low-noise fully differential voltage buffer

The well-known Sallen–Key (SK) families of filters are attractive as they utilize a single voltage buffer to implement cascadable continuous time filter sections. Most high-performance analogue integrated circuits incorporate fully differential signal paths. This is because fully differential operation improves the performance of mixed analogue/digital systems in terms of supply noise rejection, dynamic range and harmonic distortion. To achieve a high dynamic range, fully differential version of SK filters, a fully differential buffer (FDB) circuit with low noise and high linearity is required. The required fully differential buffer cannot be designed using a regular op-amp. It is necessary to find a CMOS circuit that will lead to the realisation of a fully differential buffer with low power consumption and low flicker noise.

3.2.1 Differential difference amplifier-based fully differential buffer

In this section we present a class AB CMOS fully differential buffer circuit that provides low standby current consumption, high current driving capability and low input referred noise.

The proposed fully differential buffer is achieved using a fully balanced version of the differential difference amplifier (DDA) [2, 3]. The DDA is a five-terminal device as shown in Figure 3.1a. It has two differential input ports $(V_{pp} - V_{pn})$ and $(V_{np} - V_{nn})$, rather than two single-ended inputs as is the case in the conventional op-amp.

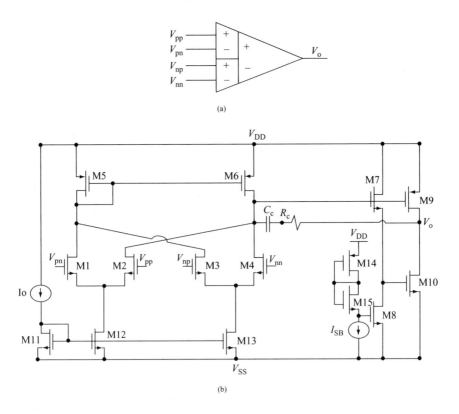

Figure 3.1   Single-ended DDA symbol: (a) symbol, (b) class AB realisation

The output of the DDA can be expressed as

$$V_{op} = A_o\big[(V_{pp} - V_{pn}) - (V_{np} - V_{nn})\big] \tag{3.1}$$

where A_o is the open-loop gain of the DDA. Analogous to a traditional op-amp, when negative feedback is applied the differential voltages of the two input ports become equal:

$$V_{pp} - V_{pn} = V_{np} - V_{nn} \quad \text{as} \quad A_o \to \infty. \tag{3.2}$$

As the finite open loop gain A_o decreases, the difference between the two differential voltages increases. Therefore, high open-loop gain is required to improve performance.

3.2.2 MOS realisation of a fully differential DDA

CMOS realisation of the DDA is shown in Figure 3.1b. Like an op-amp, the DDA mainly consists of two stages: a differential-input transconductance stage (differential amplifier with active loads) and a second gain stage (common-source amplifier with an active load). However, the DDA uses two differential pairs to implement the two input ports. For low-power operation and high current driving capabilities, a class AB output stage has been employed instead of the class A counterpart. The two differential pairs convert the two differential voltages into two currents that are subtracted, converted into a voltage by the active load, and amplified by the second stage.

A compensating capacitor (C_c) and resistor (R_c) are employed to ensure stability. A rail-to-rail low power output stage is incorporated. It consists of transistors M7–M10 and biasing transistors M14 and M15. In the absence of input signals, no current is withdrawn from the output terminal. The currents of M9 and M10 are set equal to a small standby current (I_{SB}). However, when the circuit is supplying current, M9 will be fully conducting while M10 will be almost off. Similarly, when the circuit is sinking current, the current of M10 will dominate and that of M9 will be negligible. A fully balanced DDA (FBDDA) can be designed in much the same way as a conventional op-amp.

Figure 3.2 shows the circuit diagram of the fully balanced DDA. The circuit provides two balanced outputs ($V_{op} = -V_{on}$) that are centred about a well-controlled common-mode voltage level. A CMFB circuit is needed to establish the common-mode output voltage, and without it the common-mode voltage output would drift. The CMFB circuit determines the output common-mode voltage and controls it such that it is equal to some specified voltage (usually mid-rail) even in the presence of large differential signals. When dual power supplies are used, V_{cm} is set to zero. The CMFB circuit used in Figure 3.2 consists of transistors Mc1–Mc7 in addition to two resistors (R_{cm}) and two capacitors (C_{cm}).

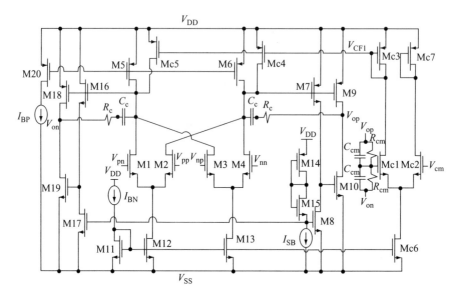

Figure 3.2   The fully differential DDA CMOS realisation

3.2.3   Wide-range low-noise FDB based on the DDA

The input differential range of the DDA is determined by the valid area of operation of the differential pairs at the input ports. It is well known that the profile of the differential output current (I_D) of a differential pair versus its differential input voltage (V_d) can be expressed as:

$$I_D = \begin{cases} -I_o & \text{if } V_d \leq -\sqrt{\dfrac{2I_o}{K}} \\[3mm] \dfrac{1}{2}KV_d\sqrt{\dfrac{4I_o}{K} - V_d^2} & \text{if } |V_d| \leq \sqrt{\dfrac{2I_o}{K}} \\[3mm] I_o & \text{if } V_d \geq \sqrt{\dfrac{2I_o}{K}} \end{cases} \qquad (3.3)$$

where I_o is the tail current and K is the transconductance parameter of MOS transistors. Operation of the circuit is valid as long as I_D is proportional to V_d, in other words, transistors carry a current or $|V_d| \leq \sqrt{(2I_o/K)}$. Outside this region, either current of the differential pair is zero. Therefore, to maintain a wide input and output swing, feedback should be applied in such a way that it forces V_d to have a small value which is very close to zero. In general, negative feedback from the outputs of the DDA to two of its inputs can be achieved in four ways as demonstrated in Figure 3.3. Since the DDA is symmetrical these four combinations can be reduced to two categories. The first category, Figure 3.3a, b, exhibits negative feedback to the same differential pair while the second, Figure 3.3c, d, comprises the feedback to both differential pairs.

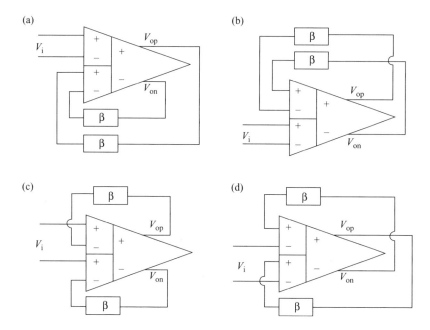

Figure 3.3   Negative feedback combinations

The first category behaves similar to the DDA since the differential pair, which has no feedback applied to it, will not exhibit a virtual short between its inputs. The second category of circuits will behave like an op-amp because the feedbacks are applied to each of the differential pairs, and hence each pair will exhibit a virtual short between its inputs. Therefore, both differential pairs will operate near $V_d = 0$, and $g_{mn}(W/L)_n$ can be designed as large as desired to achieve low noise, high input range, and low distortion simultaneously. Simulation of the DC transfer characteristics indicates that the circuit of Figure 3.3c has the widest linear input differential range for $\beta = 1$.

The other major design factor that determines the performance of the DDA-based buffer is noise. A MOS transistor typically generates two types of noise: flicker ($1/f$) noise and thermal noise [4]. It can be shown that thermal and flicker input referred noises of the proposed FDB are given by:

$$V_{thermal}^2(f) = \frac{32}{3}KT\left(\frac{1}{g_{mn}}\right) + \frac{32}{3}KT\left(\frac{g_{mp}}{g_{mn}}\right)^2\left(\frac{1}{g_{mp}}\right) \tag{3.4}$$

$$V_{flicker}^2(f) = \frac{4}{C_{ox}f}\left[\frac{K_n}{W_n L_n} + \left(\frac{\mu_p}{\mu_n}\right)\left(\frac{K_p L_n}{W_n L_p^2}\right)\right] \tag{3.5}$$

where K is Boltzmann's constant (1.38×10^{-23} J K^{-1}), T is the temperature in kelvin, $g_{mn} = g_{mi}$ ($i = 1$ to 4), $g_{mp} = g_{m5} = g_{m6} = g_{mc1} = g_{mc2}$ are the small signal transconductances of the MOS transistors, K_n and K_p are the flicker noise constant

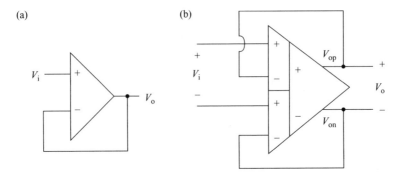

Figure 3.4   Voltage buffer: (a) single ended, (b) fully differential

for the NMOS and PMOS transistors, respectively, W and L are the width and the length of the transistors, μ_n and μ_p are the carrier mobilities, C_{ox} is the gate oxide capacitance per unit area, and f is the frequency. It is clear that the transconductance (g_{mn}) of the differential pairs should be made as large as possible to minimize thermal noise. Also, increasing the widths of the differential pair transistors clearly minimizes $1/f$ noise. Moreover, it can be shown that the flicker noise exhibits a minimum for an optimum L_n. This is because L_n appears in the numerator and the denominator of the flicker noise equation. The optimum L_n is given by:

$$L_n = \sqrt{\frac{\mu_n K_n}{\mu_p K_p}} L_p. \tag{3.6}$$

Simulation results show that the profile of the input referred noise of the DDA decreases as $(W/L)_n$ is increased. More specifically, the flicker noise decreases from 175 nV/$\sqrt{\text{Hz}}$ to 25 nV/$\sqrt{\text{Hz}}$ and the thermal noise drops from 38 nV/$\sqrt{\text{Hz}}$ to 8 nV/$\sqrt{\text{Hz}}$ as $(W/L)_n$ is increased from 9 μm/3 μm to 498 μm/3 μm. A fully differential buffer (FDB) is developed by configuring the DDA in a unity gain feedback configuration as shown in Figure 3.4a, b. DC transfer characteristics of the single-ended buffer and fully differential buffer are as shown in Figure 3.5a and b, respectively. It can be seen that the FDB exhibits a wide input differential range of 2.6 V.

3.2.4   Highly linear DDA-based filter section

The proposed filter section is based on the fully differential buffer circuit connected in a Sallen–Key lowpass filter configuration. The filter utilises one active element to realise a programmable third-order section. This results in a simple filter section that can be effectively used as a building block for implementing higher-order filters. It simplifies the implementation of various higher-order filters and shortens the design time. The buffer circuit provides class AB low-impedance buffered output for low power consumption and high-speed operation. A fully differential version of the

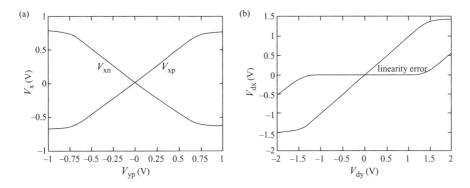

Figure 3.5   The measured DC characteristics: (a) single ended, (b) fully differential

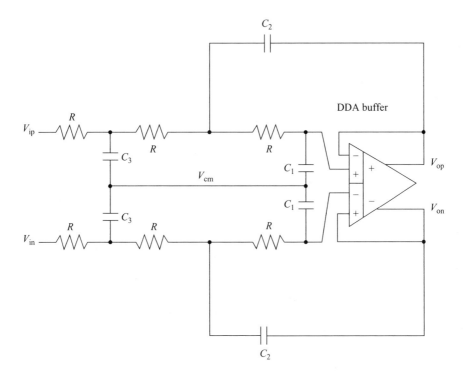

Figure 3.6   Fully differential third-order Sallen–Key lowpass filter

third-order SK lowpass filter is shown in Figure 3.6. The filter characteristics are determined by numerical values of the coefficients in the filter transfer function. SPICE simulations of the entire filter were used to determine the optimum passive component values. The bandwidth of the filter can be varied without changing the quality factor by tuning the resistors together.

Programmability is achieved by using resistor/capacitor arrays and switches to digitally select the appropriate set of components for each standard. Filter bandwidth is programmed to accommodate the GSM, IS-95 and 3G WCDMA standards. Polysilicon resistors are used to achieve high linearity and low signal distortion. This is important in integrated receivers due to the presence of large out-of-band blockers that can result in intermodulation products with frequencies inside the passband of the filter, corrupting the desired signal. This filter section is also useful in achieving non-tuneable highly linear low-noise pre-filters. Such filters are usually used in integrated receivers to attenuate the out-of-band blockers and hence relax the linearity requirements on the following parts of the baseband chain. They are also useful as anti-aliasing filters. The FDB-based filter has the following features:

(1)   simple structure with one active element per third-order filter section resulting in short design time, low power consumption and good dynamic range;
(2)   buffered voltage output that simplifies cascading;
(3)   can be implemented using regular CMOS processes (no floating wells required);
(4)   class AB operation (low standby power + high current drive capability).

The low-power fully differential filter section was fabricated in a 1.2 μm Nwell CMOS process available through MOSIS. The supply voltages were set to ± 1.5 V and the total standby current of the filter was 350 μA. The filter is designed to exhibit a bandwidth of about 1.5–2 times the bandwidth of the desired channel. This allows a safety margin to the passband characteristics of the filter against components and temperature variations. Figure 3.7 shows the measured lowpass magnitude responses covering the GSM, IS-95 and WCDMA standards with bandwidths of 100 kHz, 700 kHz and 2 MHz, respectively. The filter section provides a digitally programmable passband that provides a passband ripple of less than 0.2 dB while providing an attenuation of about 70 dB for the blockers. The non-linearity of the filter is measured by the third-order intermodulation distortion. Two sinusoidal signals were applied to the input and the output third-order intermodulation was measured for different input amplitudes. Figure 3.8 shows a typical signal spectrum of the third-order intermodulation distortion. The wide range highly linear performance of the filter section is evident from this figure, as the intermodulation components are more than 66 dB below the blocker signal level. It is also worth mentioning that this is achieved without any attenuation of the two blocking signals since they were placed in the pass band of the filter section.

3.2.5   Tuneable filter section

The filter section shown in Figure 3.6 offers no frequency tuning. The DDA-based filter of Figure 3.6 can be modified to build MOS-C tuneable filters. This can be accomplished by placing MOS transistors in parallel with the resistors used in realising the fully differential Sallen–Key sections as shown in Figure 3.9. The fully differential nature of the filter cancels the even-order harmonics.

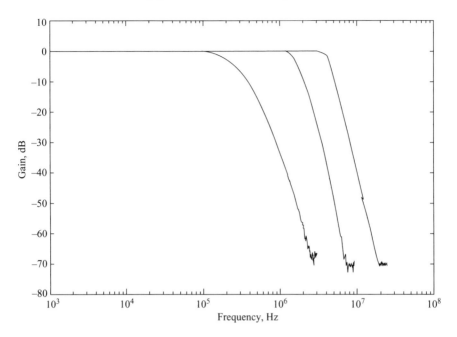

Figure 3.7 AC response of SK third-order filter

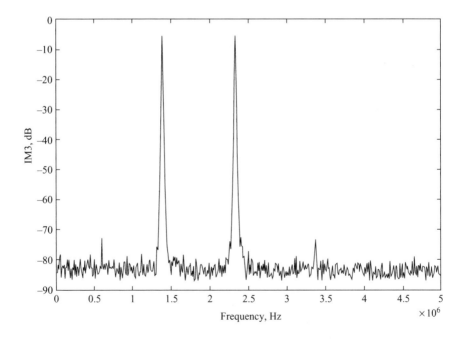

Figure 3.8 In-band IM3 of SK third-order filter

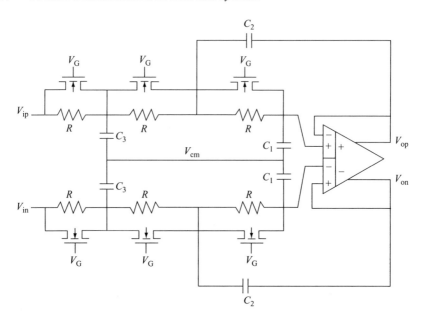

Figure 3.9   Fully differential third-order DDA MOS-C Sallen–Key filter

The filter can be tuned by changing the gate voltage of the triode region MOS resistors. Although the even-order harmonics will be cancelled by the fully balanced nature of the circuit, odd harmonics are still present. While this filter is not suitable for rejecting large out-of-band signals, it is still useful when used after a highly linear pre-filter that attenuates the out-of-band blockers. The proposed filter sections are attractive because they are simple to design and employ relatively few active elements. The high-bandwidth class AB nature of the FDB makes the filter suitable for handling high-bandwidth signals with low power consumption. Figure 3.10 shows the frequency response of a sixth-order MOS-C filter.

3.2.6   Variable gain amplifier

The third important element in a reconfigurable baseband chain is the variable gain amplifier (VGA). An important requirement for the VGA circuit is to provide good linearity for a wide range of signal swings. The other important consideration in VGA design is DC offset compensation. As outlined earlier, a small DC offset can be amplified by the VGA to a level that saturates the following stages or may cause the output signal to be clipped. Thus, dynamic compensation of the DC offset is an important part of the AGC loop design. Although a voltage buffer can be used to implement filter sections, it cannot be used directly to provide gain. However, sensing the current of the output terminals of FDB and conveying it to an additional current port using current mirrors extends its use to provide gain [5, 6]. By applying

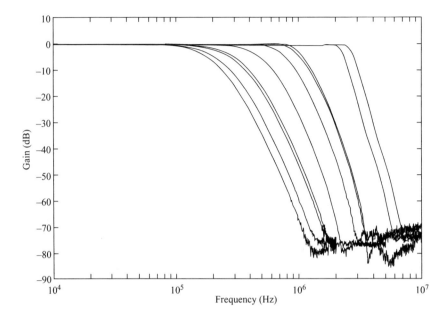

Figure 3.10   AC response of sixth-order DDA MOS-C filter

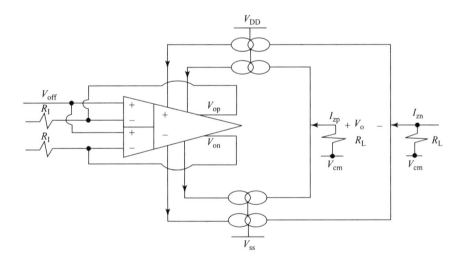

Figure 3.11   Digitally programmable VGA with DC trimming

the current sensing approach to the FDB, currents through the two output voltages are copied to two new additional current ports (I_{zp}, I_{zn}) as shown in the circuit of Figure 3.11 (current mirrors sense the currents of the voltage ports and copy them to the new current ports).

Two resistors of fixed values are connected to the input terminal while resistor arrays are connected at the output terminals to program the gain digitally. The input voltage minus the DC offset (V_{off}) is converted to a current via R_1. This current is conveyed to the output and converted back to a voltage signal by resistors connected at the output terminals. Therefore, the VGA output voltage can be expressed as:

$$V_o = \frac{R_L}{R_1}(V_i - V_{off}). \tag{3.7}$$

Thus, the gain is set by the ratio of the resistors, and the output DC offset can be cancelled using the V_{off} terminal. Note that the current-sensed FDB (CSFDB) is configured in a unity gain feedback topology. Therefore, the bandwidth of the amplifier is constant and a maximum which circumvents the finite gain bandwidth product limitation of the DDA.

Resistor arrays (R_L) are used to digitally program the gain. Figure 3.12 shows the frequency response of the variable gain amplifier. It can be seen that the amplifier exhibits almost constant bandwidth at different gain settings.

Input referred flicker noise of the VGA was measured to be 40 nV/$\sqrt{Hz}$ at 1 kHz and the thermal noise was 4 nV/$\sqrt{Hz}$ at the maximum gain setting. At the minimum gain setting, they were 61 nV/$\sqrt{Hz}$ and 16 nV/$\sqrt{Hz}$, respectively. The

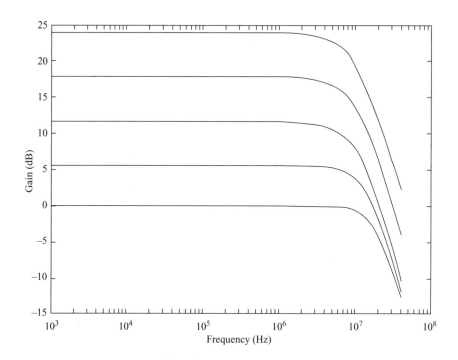

Figure 3.12   The measured AC response of the VGA

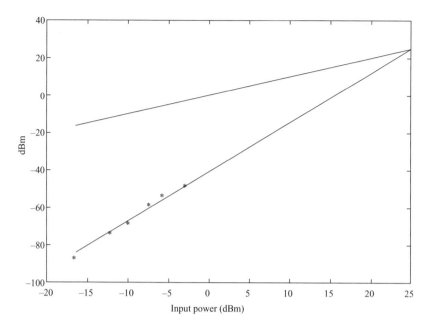

Figure 3.13   The measured IIP3 of the VGA at the maximum gain setting

Table 3.1   Measured performance of the reconfigurable chain blocks

Block	In-band IIP3 (dBm)	Out-band IIP3 (dBm)	Noise nV/$\sqrt{\text{Hz}}$
Pre-filter			
GSM	25.2	41	31
IS-95	24.1	38.4	20
WCDMA	17	25.6	17
Ch. select filter			
GSM	11.8	26	50
IS-95	15.6	20.6	30
WCDMA	16.4	13.4	26.5
VGA			
Max. gain	25	–	16
Min. gain	17	–	4.1

third-order intercept point (IIP3) of the VGA at maximum gain setting is shown in Figure 3.13. Table 3.1 summarises the performance of different baseband blocks in terms of linearity and noise performance. Figure 3.14 shows a die photo of the DDA-based filters and VGA. The two filter sections on the left can be used in a tuneable

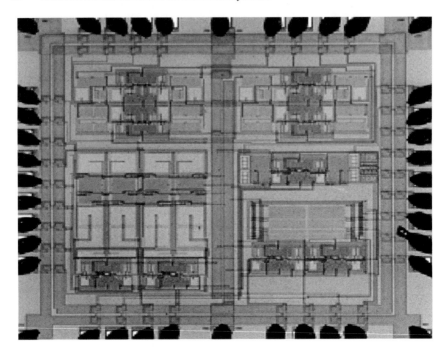

Figure 3.14   Die photo of the DDA-based programmable filter and amplifier section

MOS-C mode or a highly linear anti-alias filter (AAF) mode. The two sections on the right are VGAs based on supply sensing.

3.3   Reconfigurable baseband chain for 3G receivers

As shown in the previous sections, the DDA is a versatile element that can be used as a building block for designing filter and VGA sections. Design of a specific baseband chain can be achieved by cascading these sections. The number of sections used depends mainly on the wireless standard at hand as well as the ADC resolution. This specifies the required filter attenuation and the VGA control range. For multistandard operation, it is desirable to have a baseband chain where the number of filter and VGA sections used can be varied for each standard. The order in which the gain and filtering blocks are arranged directly affects the performance of the receiver in terms of linearity (IIP3), noise and dynamic range. Therefore a baseband chain with filter and VGA blocks that can be reconfigured in different arrangements is highly desirable. Such a chain will result in more optimized reception and will provide a programmable solution that can be easily re-configured to various demands.

To meet these requirements, the structure shown in Figure 3.15 is proposed. The baseband section consists of VGA and filter blocks that can be digitally programmed by the DSP at two levels. At the individual block level, the DSP sets the gain of the

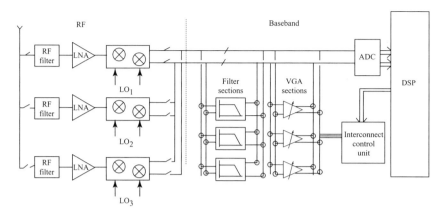

Figure 3.15   Block diagram of the reconfigurable receiver

VGA sections as well as the cut-off frequency of the filters to accommodate different wireless standards. At the architecture level, the DSP controls interconnection of the blocks. This is done by using analogue switch arrays that enable each block to be connected to the analogue bus.

For different wireless standards, the filter/VGA combination can be configured in various orders. Extra filter and VGA sections are turned off to reduce total power consumption. In each configuration the baseband chain provides a different gain control range, out-of-band signal attenuation, input referred noise and IIP3. To illustrate this, the chain is programmed in several different configurations as shown in Figure 3.16. Various parameters for each configuration are reported in Table 3.2. Here OIP3 is the output referred IP3 and SFDR is the spurious free dynamic range. In configuration (a) of Figure 3.16, the three filter sections are placed at the beginning of the chain. Two filters are used in their non-tuneable form. This is done by short-circuiting the gate of the triode region MOS transistor to analogue ground. The third filter section is tuneable and is used to reject nearby channels. The VGA sections are all cascaded after the filter sections.

In this configuration, excellent attenuation of out-of-band blockers is achieved, and the intermodulation products are strongly reduced. Furthermore, linearity of the VGA sections used is not crucial. However, cascading three filter sections without amplification decreases the signal-to-noise ratio and hence the input referred noise is expected to increase. Thus, in this configuration noise is traded off to achieve a receiver chain with high linearity and low intermodulation distortion. The chain is then programmed to be configured differently. In configuration (c) a VGA section is used at the beginning of the chain to improve overall noise performance. The VGA section is followed by a filter section to prevent further amplification of the out-of-band blockers. This filter section is subjected to large out-of-band blocker signals, and hence is configured in the non-tuneable mode to enhance linearity. For further attenuation of the blockers a tuneable filter section is cascaded next. Finally, amplification is done

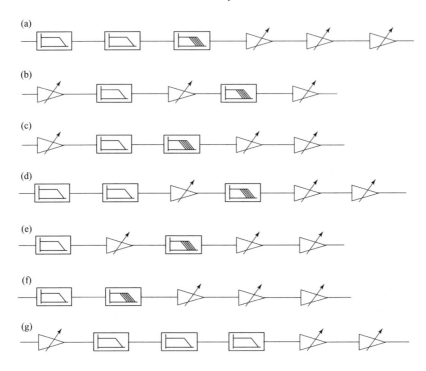

Figure 3.16   Block diagram showing the different ways in which the proposed receiver architecture can be configured

Table 3.2   Results summary for different receiver configurations

Configuration	Noise (nV/√Hz)	OIP3 (dBm)	SFDR	Power (mW)	Att. @ 1.6 MHz	Gain
(a)	22.6	29	73.4	7.7	97	48
(b)	31.88	26.7	69.8	6.5	95	48
(c)	20	26	72.5	7.7	96	48
(d)	45.6	35.3	68.2	7.7	150	54
(e)	35.1	31.5	68.3	6.5	95	48
(f)	105.2	30.7	64.1	5.4	97	48
(g)	109.5	34.2	64.6	6.5	151	54

by placing two VGA sections at the end of the chain. Various signal levels through the chain in this configuration are shown in Figure 3.17. The chain is configured to receive a GSM signal with a bandwidth of 100 kHz. Attenuation of the blocker signals at 1.6 MHz and 3 MHz is also shown in the figure. Clearly, the programmable and reconfigurable features of the proposed receiver chain will allow the designer

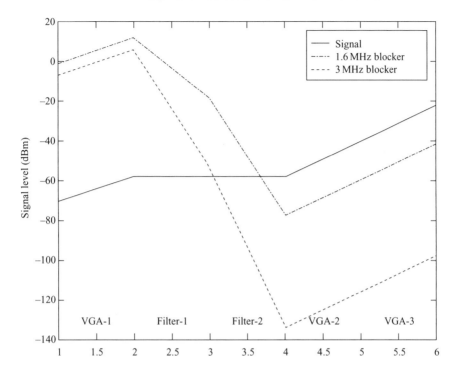

Figure 3.17   Signal level through the receiver chain

to quickly and easily trade-off between different design parameters without the need of redesigning the entire receiver chain. This approach helps in reducing the design time for the analogue portion of the wireless receivers. Furthermore, in terms of performance, the programmable nature of the chain allows a different configuration to be used for each standard, hence more optimum reception is possible.

3.4   Conclusions

The need for multimode operation makes it necessary to develop a programmable analogue baseband chain that can serve the stringent and often widely varying performance requirements of 2G and 3G wireless communication standards. To reduce the design cycle time, it would be helpful to have a single active block, which can be repeatedly used in realising required filters and variable gain amplifiers. This chapter introduces a third-order Sallen–Key filter section based on a single high-performance fully balanced differential difference amplifier based buffer. A sixth-order filter is implemented by cascading two third-order blocks. Triode region transistors in parallel with polysilicon resistors are used for tuning the filter characteristics. The same structure is modified by adding current sensing at the output of the buffer to realise a variable gain amplifier with offset control. The gain is set by the ratio of resistors and is independent of the bandwidth. Techniques to reconfigure the baseband chain for satisfying different system-level specifications constructed out of these elements are discussed. Measured results indicate that the proposed chain meets the performance specifications of GSM, IS-95 and WCDMA wireless standards.

3.5   References

1   CHEN K. C. and WU, S.: 'A programmable architecture for OFDM-CDMA', *IEEE Communications Magazine*, Nov. 1999, pp. 76–82
2   SACKINGER, E. and GUGGENBUHL, W.: 'A versatile building block: the CMOS differential difference amplifier', *IEEE Journal Solid-State Circuits*, Apr. 1987, **22**, pp. 287–94
3   HUANG, S. C., ISMAIL, M. and ZARABADI, S. R.: 'A wide range differential difference amplifier: A basic block for analog signal processing in MOS technology', *IEEE Transactions Circuits Syst.-II*, May 1993, **40**, pp. 289–301
4   JOHNS, D. and MARTIN, K.: 'Analog integrated circuit design' (John Wiley, New York, 1997)
5   TOUMAZOU, C., LIDGEY, J., and HAIGH, D.: 'Analogue IC design: The current mode approach' (Peter Peregrinus Ltd., London, 1990)
6   ELWAN, H. and SOLIMAN, A.: 'A novel CMOS current conveyor realization with an electronically tuneable current-mode filter suitable for VLSI', *IEEE Transactions Circuits Systems-I*, Sept. 1996, **43**, pp. 663–70

Chapter 4

Field-programmable and reconfigurable analogue and mixed-signal arrays

Yichuang Sun and Rushikesh Lala

4.1 Introduction

The trend in modern VLSI circuit design is to increase the level of integration, decrease the time of design and development, and reduce the cost of products. The key concept for this is to use a single hardware implementation for more than one type of system by reprogramming it for different systems in the field; this type of reconfigurability can be achieved by programmability. The need for programmability and reconfigurability has led to the use of field-programmable arrays. The advantages of these are fast prototyping, on-the-fly reconfiguration, use of PC-based design tools and reduced level of expert knowledge required to build working ICs.

Electronic and communications systems contain both analogue and digital circuits. Programmability can be easily achieved for digital circuits using field-programmable gate arrays (FPGA). The FPGA technique has been very well established. To make a whole system fully programmable and reconfigurable, the analogue part must also be programmable. Also, expensive and time-consuming analogue design needs to be automated to shorten the product-to-market cycle. This has resulted in the emergence of field-programmable analogue arrays (FPAA) [1–25]. The field-programmable mixed-signal array (FPMA) is also emerging [26], as is the system on a programmable chip (SoPC) [27].

Analogue signal processing [28–36] is normally preferred in high-speed and low-power applications. For many signal-processing applications, analogue ICs require a smaller chip area and have lower power consumption than their digital counterparts, mainly due to the lack of requirements for anti-aliasing and post-smoothing filters and A/D and D/A signal converters. For these reasons, analogue circuitry is advantageous for wireless and portable applications where compactness and low-power consumption are important. Also, analogue signal processing circuits can work at much higher frequencies than digital equivalents, and thus in high bandwidth applications such as wireless communication transceivers and computer hard disk read/write channels, analogue circuits have been widely used. FPAAs, which offer rapid prototyping on a single IC, are proving to be a very useful tool for the design of working analogue and mixed-signal ICs. In many cases, the complexity of digital circuit design can also be reduced by using programmable analogue arrays for the analogue part in the system [2].

As is generally the case for analogue against digital design, there are fundamental difficulties in realising universal FPAAs over FPGAs. Apart from those well-known challenges such as linearity, noise and bandwidth subject to process and environmental variations, the configurable analogue blocks in the FPAA must provide a number of programmable functions and the configurable routing should not decrease the accuracy of the implemented circuit. Research into FPAAs and applications has attracted substantial interest across the whole IC sector including both academia and industry. Several FPAA products have been commercially available from companies [21–25] for research, development and teaching uses and continued activities in the field are also actively being conducted.

This chapter overviews FPAA. General concepts, architectures, design issues, circuit techniques and applications of FPAAs are presented. Commercial FPAAs are reviewed and a CMOS OTA-C FPAA is described. The concepts of FPMA and systems on a programmable chip are also introduced. Potential applications of FPAAs and FPMAs in wireless communications are discussed. The chapter is organised in the following way. Section 4.2 is concerned with generic concepts, architectures, design issues, advanced circuit techniques, and general applications of FPAAs. Section 4.3 overviews commercial FPAA products with different bandwidths and circuit techniques. The design and implementation of a continuous-time high-frequency CMOS OTA-C FPAA is then presented in Section 4.4. Section 4.5 introduces the concepts of FPMA and systems on a programmable chip (SoPC). Potential applications of FPAA and FPMA in universal transceivers are described in Section 4.6. Finally, Section 4.7 concludes the chapter with some future research perspectives of the FPAA and FPMA.

4.2   Field-programmable analogue arrays (FPAA)

An integrated circuit that can be programmed to implement analogue circuits using flexible analogue blocks and interconnection is known as an FPAA. Both circuit configuration and parameters are changeable using on-chip memories [4].

4.2.1   General architecture

Field-programmable analogue arrays contain an array of programmable function blocks (FB) or configurable analogue blocks (CAB), which can be reconfigured or reprogrammed for different analogue signal processing functions with the aid of the programmable interconnection network. Both the function block and interconnection

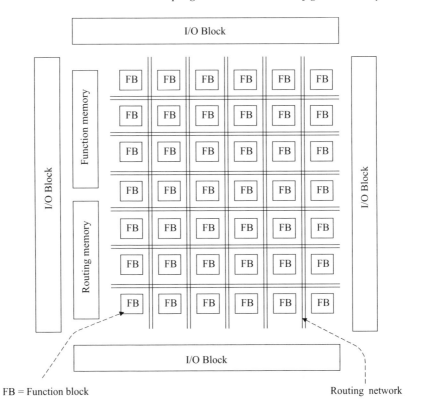

Figure 4.1   *Generic FPAA architecture*

network can be programmed by on-chip memory. The specially designed software for particular FPAA chips enables the user to design and simulate the analogue applications and then send/write the data to the FPAA hardware to program the device. The different chip's hardware is designed to facilitate the electrical evaluation of the design. The generic block diagram of an FPAA is shown in Figure 4.1 [3, 7, 21]. The routing memory stores a bit string which is used for the connectivity of the intercon- nection network, and the function memory stores a bit string which is used for function generation of the FB/CABs and fine tuning of function parameters. Interconnection networks can take the form of a tree, crossbar or datapath, which connect the FB/CABs and input/output signals for different given system requirements. Each FB/CAB can be programmed to implement some basic analogue functions such as adders, multi- pliers, amplifiers, integrators, etc. FB/CABs can be designed for both specialised and general-purpose FPAAs. The performance parameters such as the gain, frequency and Q are tuneable through controlling the transconductances/MOSFETs and capacitors in continuous ranges. Many on-chip tuning methods such as master–slave, PLL and adaptive tuning, can be used [37–39].

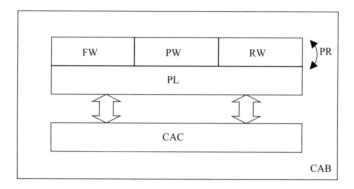

Figure 4.2   Configurable analogue block

4.2.2   Configurable analogue blocks (CABs)

The CAB is the main function block in FPAAs, and it can be configured to perform different analogue signal processing functions. Each of the CABs has three subcells: a programming register (PR), programming logic (PL), and a configurable analogue cell (CAC). The CAC's functions can be programmed digitally using a function word (FW). The parameter word (PW) can be used to program the characteristics (parameters) of the configured function. The desired connectivity between the CACs in different CABs is configured through the routing word (RW). The digital control word of each CAC is held in the respective PR. The entire PR makes a single shift register which can be loaded with the desired programming bit stream through one external pin of the FPAA. The PL carries a predefined mode of operations for different CAC configurations and routing networks which are configurable by the PR. The I/O blocks are arranged surrounding all the four sides of the array to provide the input and output interface for the chip. Figure 4.2 shows the block diagram of a CAB [7]. A specialised CAB may have a few different functions, whilst a general CAB can have many functions. General CABs are normally more complex than specialised CABs. CABs can be designed at transistor level or building block level. Building block level design using op-amps, OTAs, etc. is preferred in terms of chip size [2].

4.2.3   CAD design procedure using FPAAs

FPAAs can be configured for different applications by configuring the CAB and interconnections with the help of CAD (computer-aided design) software. All commercial FPAA chips can be easily prototyped and reprogrammed using the software available with them. The commercially available FPAAs also have built-in IP (intellectual property) modules available in their software so that users can easily use the ready-made functions for their own design. The flowchart illustrated in Figure 4.3 [2] provides the CAD procedure, which can be used to implement an analogue circuit on to an FPAA. After finishing pre-study, top-level design and module specification of the project, the analogue module is simulated in the module implementation phase. A GUI-based

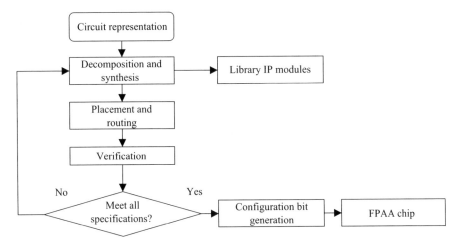

Figure 4.3   CAD methodology using FPAA

schematic editor is used to generate a netlist from the desired analogue circuit. The schematic can be of many levels from behaviour down to transistor level depending on the synthesising capacity of the CAD tool. Based on the resources available on the FPAA chip, circuit simulation and synthesis can be done using the CAD tool. Then the circuit is placed and routed, and the original schematic is back annotated. Verification is performed to check if all design specifications are met. If not, the whole procedure can be repeated. When design specifications are met, a configuration bit string is generated by the CAD tool and downloaded onto the FPAA chip. The downloading is usually performed using the port of a computer; often the configuration can also be stored in an on-board EEPROM. If needed, a circuit change can be done in a very short time compared with the months it would take to redesign and fabricate a new ASIC [2].

4.2.4   General design issues for FPAA

The design of a CAB/FPAA is possible in both continuous-time and sampled-data domains. A sampled-data design uses switched-capacitor or switched-current techniques, and continuous-time design uses transconductors or op-amps. The continuous-time design has advantages in terms of bandwidth, power and chip area, but it has narrower ranges of programming parameters and is sensitive to device non-idealities. On the other hand, the sampled-data design has advantages in terms of programmability and insensitivity to device non-idealities, but it is limited to low-frequency operation by the sampling frequency and often consumes more power and occupies a larger chip area due to the extra anti-aliasing and smoothing filters and sample and hold circuits. Technically, four-quadrant multipliers are easy to implement in continuous-time, but difficult in sampled-data domain. Also, noise due to clock signals is often a problem for sampled-data systems, but it does not exist for continuous-time systems. As widely tuneable active devices and on-chip tuning circuits are used, parameter ranges and performance accuracy have been enhanced. In modern FPAAs, continuous-time techniques are more often used and will become dominant for future high-frequency applications.

Other practical issues important for designing an FPAA include switching element parasitics, choice of voltage mode or current mode, configuration storage, performance specification, etc. The switching elements are used for interconnection of CABs. Non-idealities from switches such as the voltage drop across them, on-resistance, non-linearity and parasitic capacitances are detrimental to the performance of the FPAA. These can be reduced by using subthreshold operation switches [4], transconductor switches [5] and buffered pass switches [3]. The design is possible in both voltage and current mode. Voltage-mode circuits use voltage signals, and current-mode uses current signals in operation. Typical examples of the former are continuous-time circuits using voltage amplifiers such as the op-amp-RC and MOSFET-C, and sampled-data switched-capacitor circuits; and for the latter, continuous-time circuits using current amplifiers such as current conveyer-based and log domain, and sampled-data switched-current circuits. Continuous-time OTA-C circuits can work in both voltage and current mode. Low power/voltage design is sometimes difficult in voltage mode, but it may be easily possible using current-mode techniques. Also, current-mode design may achieve higher bandwidth and larger signal swing.

4.2.5   Advanced circuit techniques for FPAA

The switched-capacitor (SC) technique is familiar to designers, and early FPAAs used this technique [11]. The op-amp-RC approach is another well-established method which has also been utilised in several FPAA designs [3–5]. More recent analogue design techniques are the switched-current (SI), log domain, MOSFET-C and OTA-C [35].

Switched-current circuits are suitable for low-cost or wideband applications in digital CMOS implementations compared with switched-capacitor circuits. Comparisons of switched-current filters and traditional switched-capacitor filters have been made, showing the advantages of the SI filters [16, 40]. SI circuits have been successfully used for video frequency signal processing. Further increase in the operating frequency may be difficult due to the sampling requirement.

Log domain circuits are based on the exponential function of bipolar or MOS transistors. Rather than linearising the transistor, this method directly uses the non-linear characteristic to synthesise linear circuits [17, 41]. The translinear and companding techniques are related [8, 17]. Recent effort has been put into implementation of log domain circuits using mainstream CMOS technology. Log domain filters are perhaps the most revolutionary development in the recent rapid development of filter theory. The frequency range of log domain CMOS circuits is restricted to a few MHz range, although bipolar circuits can be somewhat higher.

MOSFET-C filters result directly from active-RC filters with the resistor being replaced by the tuneable MOSFET [42–45]. One advantage is that the wealth of knowledge of active-RC design can be used directly. Use of balanced structures is particularly important for MOSFET-C circuit design. CMOS has been almost exclusively used for MOSFET-C filters in the literature with operation frequency typically in the MHz range. Recent research has shown that with MOSFET resistors and bipolar op-amps, BiCMOS can also be an attractive technology for high-frequency high dynamic range MOSFET-C filters [42]. An example of 120 MHz filters with an extended tuning range has been reported [43].

Whilst active-RC and MOSFET-C circuits utilise the high gain operational amplifier in closed loop form, OTA/g_m-C circuits [46–55], which are dominant in high-frequency applications, use the tuneable transconductance amplifier in open loop form and do not use (MOSFET) resistors. Both voltage-mode and current-mode g_m-C filters are available [34, 49, 50]. Automatic tuning is particularly important for g_m-C circuits, as they are sensitive to parasitics. The dominant IC technology for g_m-C circuits is CMOS, although BiCMOS, bipolar and GaAs have also been used. The CMOS g_m-C technique is a most outstanding feature of modern analogue IC design. CMOS g_m-C filters for the GHz range are emerging.

In modern analogue circuit design two key techniques have been widely used: the balanced structure and on-chip automatic tuning. Balanced architectures should be utilised for all kinds of circuits to eliminate even-order harmonic distortion and reduce coupling of noise from the power supply and digital circuits [32]. On-chip automatic tuning is crucial for fully integrated continuous-time filter design to overcome the effects of parasitics, temperature and environment changes [37]. Both frequency tuning and Q tuning may be needed.

Continuous-time ICs feature modern analogue circuit design for high bandwidth, low power and small chip applications such as computer hard disk drives [56, 57] and wireless communication transceivers [58–60]. The g_m-C approach has overshadowed all the other competing methods for high-frequency applications and will be the main technique for some time in the future. Many IC technologies have been used in practical implementations, but CMOS has been the most popular due to the low cost and suitability for single-chip systems such as single-chip transceivers for Bluetooth, and will be so for some years to come.

4.2.6   FPAA applications

As described above the FPAA has a matrix of CAB, surrounded by a programmable interconnection and I/O structure. Common analogue signal processing functions such as offset removal, rectifiers, gain stages, comparators and first-order filters can be implemented using just one CAB. More complex functions such as high-order filters, oscillators, pulse-width modulators and equalisers can be implemented using two or more CABs. The circuit's configuration is held in on-chip memory, which is initialised on power-up from EPROM, or through the chip's microprocessor peripheral interface. General-purpose FPAAs are for general signal processing and can be used to prototype and design any analogue circuits or as many as possible [7, 25].

One particular application of the FPAA is in filter design. In recent years, the renewed interest in analogue, mixed-signal and RF circuits due to system-on-chip design has led to new interest in research into high-frequency integrated analogue filters [32–36]. In fact, the high-frequency integrated analogue filter has become a key component in achieving ubiquitous communication and computing. For example, RF bandpass filter design with high Q is a challenge for single-chip transceivers in wireless communications. Linear phase filters/equalisers at higher frequencies for mass storage systems such as computer hard disk drives are an active topic.

The FPAA's application range spans most areas of electronic design, including data acquisition and control, automation, instrumentation and telecommunications. In these applications it will replace large areas of PCBs with a single component offering repeatable, drift-free performance and may save up to 90 per cent development time [23–25]. Application-specific FPAAs have been developed for some particular applications such as neural networks [1] and signal conditioning/monitoring. FPAAs can also be embedded in a system as part of the chip. For example, in almost all electronic systems containing analogue and digital circuits, there is a lowpass filter (LPF) for anti-aliasing filtering and a variable gain amplifier (VGA) for amplification and for relaxing the requirements for A/D converters. The combination of a LPF and a VGA can be seen in video signal processing systems, computer hard disk drive systems, instrumentation and control systems, and communication systems. This kind of so-called analogue frontend circuit (LPF and VGA) may be typically implemented using an FPAA. Potential applications of FPAAs in ubiquitous wireless transceivers will be discussed in detail in Section 4.6.

4.3   Commercial FPAA products

Realising their significance and market, industry has quickly come into the FPAA business. Though the programmable analogue concept has been around for only about ten years [1], several FPAA products have already been commercially available from different companies. With the current commercial and research trends in the area of reconfigurable analogue devices, more and more high-performance FPAAs will certainly appear in the future.

Some typical commercial FPAAs are presented in Table 4.1 with some comparative information about different types in terms of bandwidth, circuit techniques, IC technology and applications. As examples, we will illustrate three FPAA products from Lattice, Zetex and Anadigm, respectively, in the following sections.

4.3.1   Lattice semiconductor's programmable analogue circuits (ispPAC)

Lattice has released commercial FPAA-like op-amp-RC products. Different ispPAC family products are available for different kinds of application like lowpass filters,

Table 4.1   Commercial FPAAs

Company	Bandwidth	Circuit technique	IC technology	Applications
Motorola MPAA [13, 14, 21]	200 kHz	Switched-capacitor	CMOS	General signal processing
IMP EPAC [22]	125 kHz [2], 150 kHz [12]	Switched-capacitor	CMOS	Signal conditioning or monitoring
Lattice ispPAC [23]	1.5 MHz	Continuous-time (op-amp-RC)	CMOS	Signal processing and conditioning
Zetex TRAC [24]	4–12 MHz	Continuous-time (log domain)	Bipolar	General signal processing
Anadigm FPAA [25]	10 MHz	Switched-capacitor	CMOS	Signal processing and conditioning

analogue frontends, signal conditioning, and control loop applications. The block diagram of ispPAC20 is shown in Figure 4.4 [23], all of the ispPAC family devices having almost the same basic blocks as ispPAC20.

The ispPAC device has a number of programmable IAs (instrumentation amplifiers), OPs (operational amplifiers), and CPs (comparators). All of them can be configured to connect existing circuits. IspPAC devices also come with 8-bit multiplexed DACs, and different ranges of reference voltage are also available. Different ispPAC devices have different bandwidths dependent on application criteria. The maximum available bandwidth is 1.5 MHz in ispPAC30. All ispPAC family products can be reconfigured on the fly, by exchanging configuration bits between the SDRAM configuration shift register and an on-chip memory.

4.3.2   Zetex semiconductor's TRAC (totally reconfigurable analogue circuit)

Fast Analogue Solutions (FAS) of Zetex has introduced the totally reconfigurable analogue circuit (TRAC), a continuous-time, log-domain bipolar design operating at up to 12 MHz. The TRAC includes 20 CABs, organised in two rows of 10 CABs, each capable of implementing one of the following functions: log, anti-log, non-inverting pass, addition, negating pass, op-amp, and half-wave rectification. The interconnection network is hard-wired, as shown in Figure 4.5 [17, 24]. Topological programming is implemented by tuning CABs off, and by external wiring of the pins. By tuning a CAB off, its inputs and outputs are electrically disconnected, allowing the designer to use the output as an input to the subsequent CAB. Amplifier gain is determined by using off-chip resistors. Configuration of the CABs is accomplished using a 60-bit string. Once more, a CAD tool is used to configure the TRAC; the CAD tool includes a simulator to simulate a circuit before

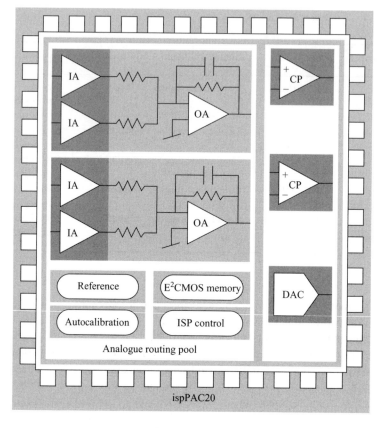

Figure 4.4   Block diagram of ispPAC20

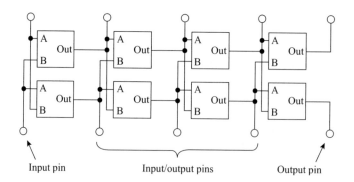

Figure 4.5   Part of FAS TRAC020 array architecture

being downloaded on to the FPAA IC. FAS advocates a computational approach to analogue circuit design, where circuits are designed with their functionality in mind (a top-down approach), rather than the underlying circuitry (a bottom-up approach).

4.3.3 Anadigm field programmable analogue array (AN10E40)

Anadigm has released a CMOS switched-capacitor FPAA design called AN10E40. This FPAA architecture, organised as an array of CABs, is illustrated in Figure 4.6 [25]. The AN10E40 consists of a 4×5 matrix of fully configurable switched-capacitor cells, enmeshed in a fabric of programmable interconnect resources. These programmable features are directed by an on-chip SRAM configuration memory. The SRAM configuration memory is initialised on power-up via an off-chip serial PROM or through a microprocessor peripheral interface. The SRAM block that controls routing connections and CAB behaviour is loaded during configuration. Configuration typically occurs at power up as an automatic process but can of course be re-initiated at any time. The ability to re-configure the part at any time gives the user flexibility in system design. Programmable capacitor banks and local switching in both the input paths to the op-amp and a programmable capacitor bank in the op-amp's feedback path provide all the resources required to implement a very large number of analogue processing circuits. AN10E40 comes with configurable clocks which can be used up to 40 MHz, although the switched capacitors and switches allow this FPAA to work

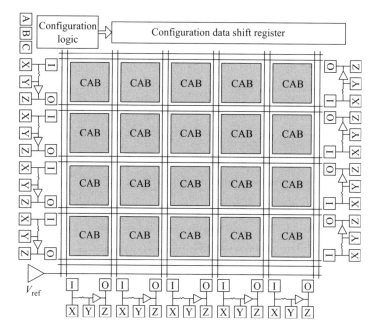

Figure 4.6   Block level view of the AN10E40 array

up to 10 MHz bandwidth. AN10E40 comes with PC-based CAD tools and quite a few ready-made ipMODUAL libraries [25].

4.4 Continuous-time CMOS OTA-C FPAA for high frequencies

For low-frequency analogue frontends in signal processing systems, the FPAA using the continuous-time MOSFET-C approach can achieve very good performance. But for high-frequency operations, the OTA-C technique is the best choice. Using the OTA with programmable transconductance (g_m), and the programmable capacitor array, it is possible to build different circuits for various high-frequency applications.

4.4.1 Programmable OTA

The simplified schematic diagram of the OTA based on two cross-coupled differential MOS pairs and digitally programmable current mirrors is shown in Figure 4.7 [18, 19]. Using the standard square-law model for MOS devices and assuming that the current gains of the programmable current mirrors are all equal to A, the output current can be derived as

$$I_{out} = \tfrac{1}{2}A(I_{in1} - I_{in2}) = K(V_{in+} - V_{in-})V_b A = K V_b A V_{id} = g_m V_{id} \qquad (4.1)$$

where $K = 0.5\mu C_{ox} W/L$ is the transconductance parameter of transistors M1–M4, having the same W/L ratios, V_b is the voltage of the floating DC source, $V_{id} = V_{in+} - V_{in-}$ is the differential input voltage and g_m is the overall transconductance of the OTA circuit.

In order to be suitable for a wide range of OTA-C filter designs, it is necessary that the transconductance of the OTA should be adjustable over a wide range. From (4.1)

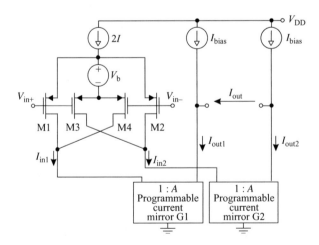

Figure 4.7   Simplified schematic diagram of the CMOS programmable OTA

it can be seen that two methods can be used to change the transconductance g_m. One method is to tune the floating bias voltage V_b by an analogue voltage V_{ctrl}. The other is to make the gain A of the output current mirrors programmable in a digital way.

A current mirror of programmable gain can be realised using the well-known high-compliance current mirror structure, using 31 identical output stages as shown in Figure 4.8 [19]. The output current I_{out} is the sum of the currents flowing through individual output stages. As can be seen from Figure 4.8, the output stages are connected in five groups of 1, 2, 4, 8 and 16 stages that can be simultaneously switched on or off by the appropriate switch S_i where $i = \{1, 2, 4, 8, 16\}$. Using n stages, the output current I_{out} can be set to a value between $I_{out}^{min} = I_{in}$ and $I_{out}^{max} = nI_{in}$ with resolution $\Delta I_{out} = I_{in}$. Only five output groups (five bits) were implemented due to the huge number of MOS devices required for the most significant bits.

In practice, every switch S_i must be accompanied by another switch $\overline{S_i}$, which can short-circuit the gate of the transistor Mia to ground. This is necessary for discharging the parasitic capacitance C_{gs} of the transistor Mia and stopping the current I_i flowing. The cascaded current mirror structure is used to achieve high output resistance.

The complete circuit diagram of the OTA including the floating voltage source and common mode feedback circuit can be found in [19]. Simulation and chip test results in CMOS show that the OTA has a transconductance tuneable/programmable in a wide range of 700 times using both V_{ctrl} and digital tuning and a 3 dB bandwidth larger than 20 MHz.

4.4.2 Programmable capacitor array

The structure of the programmable capacitor array is shown in Figure 4.9 [19]. It consists of five capacitors C_0–C_4 and switches S_{C0}–S_{C4}. The branch of capacitor C_0 and switches S_{C0} represents the least significant bit (LSB). Each branch is built of an appropriate number of LSB branches (connected in parallel).

Switches are made using MOSFETs of width high enough to achieve the phase of a capacitor in the range $-90° \pm 1°$ for frequencies up to 10 MHz. The equivalent capacitance of the capacitor array can be expressed as

$$C_{EQ} = \sum_{n=0}^{4} b_n 2^n (C_{ON} - C_{OFF}) + C_{PAR}$$

where $b_n \in \{0, 1\}$, b_n is equal to 1 when switch S_{Cn} is on and equal to 0 when S_{Cn} is off, C_{ON} and C_{OFF} are the capacitances of the LSB capacitor when switches S_{C0} are in ON and OFF modes, respectively, while C_{PAR} is the parasitic capacitance of connections when all the switches are off.

4.4.3 Configurable analogue block using programmable OTAs and capacitors

Figure 4.10 [19] shows a versatile CAB consisting of one programmable fully differential OTA, one programmable capacitor C_{EQ} and a set of switches S_1–S_{12}. The

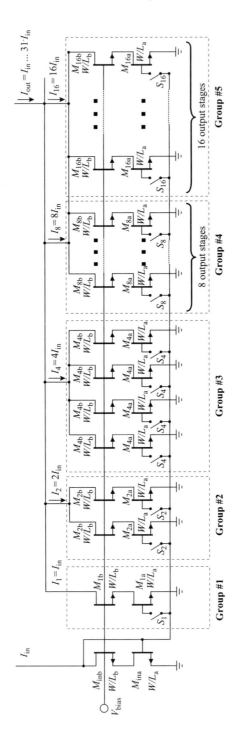

Figure 4.8   Simplified diagram of programmable current mirror array

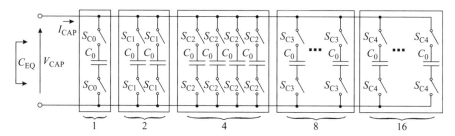

Figure 4.9   Programmable capacitor array

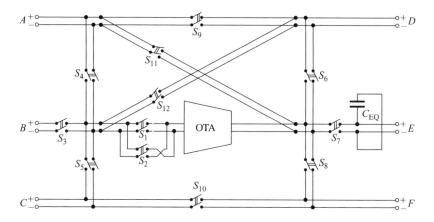

Figure 4.10   Structure of OTA-C CAB

switches are placed in such a way that the OTA can be connected with or without the capacitor. It is also possible to pass the signals of other CABs through the switches situated at the top and the bottom of the CAB. Switches S_1 and S_2 allow connection of the input to the OTA directly or in an inverse way. To program the CAB, a control word of length 22 bits must be specified (OTA: 5 bits; capacitor: 5 bits; switches: 12 bits), as well as the voltage V_{ctrl} controlling the transconductance of the OTA through V_b. The control word is stored in a shift register, while the voltage V_{ctrl} is supplied from the external automatic tuning circuit. This CAB can be configured to perform functions like addition, subtraction, amplification, attenuation, integration and filtering of signals of frequencies from several kHz up to a few MHz.

4.4.4   Fine tuning of parameters

The proposed FPAA includes three inputs and three outputs. This enables the implementation of up to three independent filters or other circuits working concurrently. Tuning of filters is based on small parameter differences between components inside the chip, so setting the reference integrator (or other reference circuit containing OTAs and Cs) to the desired g_m/C ratio will reflect on the filter characteristic frequency.

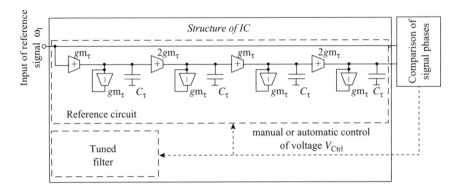

Figure 4.11   Simplified schematic diagram of the tuning circuit

The accuracy of such tuning is dependent on how precisely the desired g_m/C ratio can be achieved and the technology used. Practical implementations provide 0.1–5 per cent accuracy [19].

Figure 4.11 presents the block diagram of the tuning circuitry used during testing of the chip [19]. The upper part is the reference circuit consisting of four lossy integrators. The lower part is a tuned filter. All OTAs in the reference part as well as in the filter are identical and have the same value of control voltage V_{ctrl}. Comparison of an integrator's phase delay with its desired value gives a signal which can be used in the manual or automatic correction of V_{ctrl}.

4.4.5   FPAA using CABs and its application in OTA-C filter design

With the above building blocks and discussions, we can now proceed with the design and implementation of FPAAs. We will present an FPAA with 40 CABs and demonstrate an application example of a high-order bandpass filter.

4.4.5.1   FPAA using 40 CABs

The FPAA used for implementation of filters is presented in Figure 4.12 (block diagram) and Figure 4.13 (die photograph) [19]. It consists of 40 CABs from Figure 4.10 positioned in eight columns and five rows. Additionally three OTAs o1–o3 act as signal buffers. Input signals are delivered through lines i1, i2 and i3. Because of this, up to three different filters can be implemented simultaneously. The transconductance parameters of all the OTAs are controlled by the external voltage V_{ctrl} and through digital switching of the output current mirrors. While the voltage V_{ctrl} is common for all the amplifiers in the array, it is still possible to set the transconductance of every OTA separately by setting the gain of the OTA's current mirror. The FPAA was physically implemented in the CMOS process through MOSIS. Programming of the FPAA is performed through serially shifting digital words of 880 bits into programming registers.

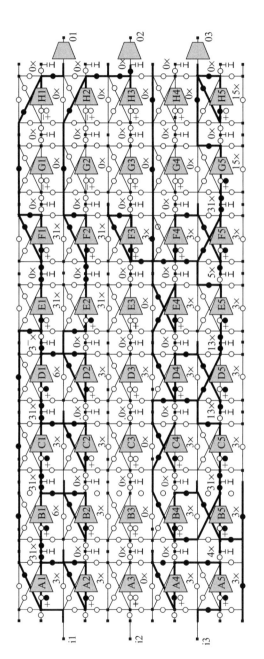

Figure 4.12   Structure of OTA-C FPAA. Bold lines represent active connection, which realises a sixth-order bandpass filter with tuning circuitry

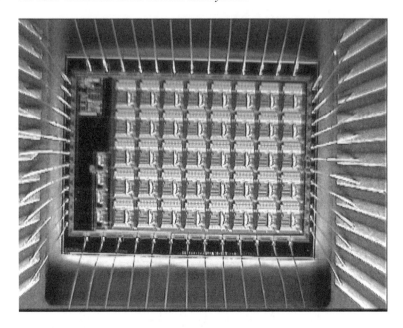

Figure 4.13   Layout of matrix of 5 × 8 CABs in Figure 4.12

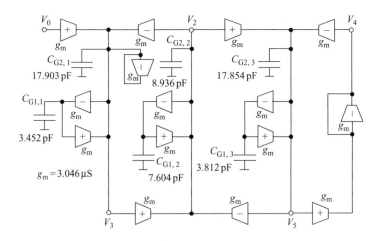

Figure 4.14   OTA-C realisation of sixth-order bandpass filter

4.4.5.2   Implementation of a leap-frog bandpass OTA-C filter using FPAA

A sixth-order bandpass filter with centre frequency $f_0 = 60\,\text{kHz}$ and passband $\text{BW} = f_0$ is designed based on simulation of a passive LCR ladder [19]. The schematic diagram of the filter is presented in Figure 4.14. The placement of Figure 4.14 in the FPAA resources is shown in Figure 4.12. The calculated values have been rounded to the nearest values realisable in the FPAA. The capacitor's values also include the parasitic capacitances of switches and connections. Adjusting the analogue voltage V_{ctrl} can change the centre frequency of the filter about 22 times. In Figure 4.12, the upper part (i.e. the first and second row of CABs: A1–H1 and A2–H2) includes the tuning circuitry described in Section 4.4.4. Additionally, in the practically implemented tuning circuit in Figure 4.12, OTAs E1 and E2 have opposite transconductance values, so the signals actually observed at the outputs o1 and o2 should have exactly the same phase (not $-180°$). The frequency f_{i1} of the reference input signal can be calculated.

The filter was implemented in the structure of the FPAA. Measurements were taken after applying a signal of frequency f_{i1} to the input i1 (Figure 4.12) and manually setting the value of the control voltage V_{ctrl} so that signals observed at the outputs o1 and o2 were in exactly the same phase. This tuning technique was described in Section 4.4.4. Comparing with the ideal characteristic, the filter exhibits no more than 1 dB error in the range 2–200 kHz. Switches placed in the paths conducting relatively large currents are the main cause of this error.

4.5   FPMA and SoPC

Today's electronic systems are very complex and include digital logic, interfacing as well as analogue circuits. Eventually, such mixed-signal systems will be integrated on a single programmable chip. This section introduces the relevant FPMA, SoC and SoPC concepts and systems.

4.5.1   FPMA

An FPMA is an integrated circuit containing both an FPGA and an FPAA, connected through configurable signal converters as shown in Figure 4.15. The FPMA is a hybrid chip combining analogue and digital cells. The digital cells are standard FPGA circuits, while the analogue cells constitute a standard FPAA [26].

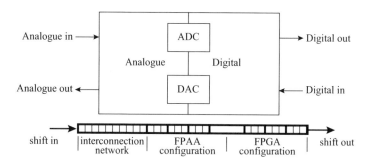

Figure 4.15   Generic FPMA

4.5.2   SoC and SoPC

The concept of systems on a single chip is not new and for years researchers have been trying to implement a whole application on a single piece of silicon. The implementation of different parts of the application on a single chip is known as system on a chip (SoC). There are quite a few reasons for implementing the SoC concept. These include: increased system performance, reduced system power dissipation, reduced product cost, smaller system size, and system consideration. The disadvantages of SoC are higher development cost and time. This could be solved by introducing the SoPC (system on a programmable chip). The concept of SoPC is developing a complete solution for any kind of system. The SoPC may contain both the FPAA and FPGA, which is similar to the FPMA. The future reconfigurable universal transceiver should ideally be a SoPC device. But unfortunately to implement this kind of universal SoPC with CMOS only is very difficult and there have hardly been any commercial products available. The SoPC from Altera is restricted to digital circuits only [27].

4.5.3   Commercial mixed-signal SoPC: SIDSA FIPSOC

FIPSOC (FIeld-Programmable System-On-a-Chip) provides the user with the possibility of integrating a microprocessor core along with programmable digital and analogue cells within the same integrated circuit [26]. This chip can be considered as a large granularity FPGA (logic cells) with a FPAA (analogue cells) and a built-in microprocessor core (8051) that not only acts as a general-purpose processing element, but also configures the programmable cells and their interconnections. Therefore, there is an interaction between hardware and software as long as signal values and configuration data within the programmable cells are accessible from the element of the FPGA. The block diagram of the FIPSOC is shown in Figure 4.16 [26]. The FPAA is for signal conditioning and uses the continuous-time circuit techniques. Configurable converter blocks produce 8–10 bits resolution.

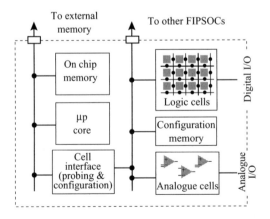

Figure 4.16   Block diagram of FIPSOC

4.6   Reconfigurable universal transceivers using FPAAs

Universal transceivers are required to meet various wireless communication standards. Universality requires that the transceivers should be programmable and reconfigurable. FPAAs may be used to achieve different functions, for example, different bandwidths, in future ubiquitous transceivers. This section addresses these issues.

4.6.1   Universal transceivers for different standards

FPAAs can be used for development or prototyping purposes. They can also be used in many other areas as mentioned in Section 4.2.6. One example may be to use FPAAs in multistandard transceivers [61–65]. Due to its programmability and reconfigurability, using the FPAA in the analogue baseband or IF stage or even RF frontend, it is possible to obtain universal transceivers that support all the standards for mobile communications in the world. The expanding growth of wireless communications has led to the proliferation of different standards. The market demands low-cost, low-power and small form-factor devices. Different second generation (2G) mobile systems have effectively utilised digital technology to increase capacity, improve reliability and lower system size and cost. However, they were developed to handle voice and low data rates and are not capable of handling traffic of high data rates like multimedia applications. Thus third generation (3G) wireless systems have emerged, which aim to integrate multiple applications, combined from consumer electronics, computer and communication markets. Applications such as audio/video, interactive video services, telephoning, e-mail, and internet access with real-time image transfer are expected. To achieve compatibility and global roaming, 3G-enabled transceivers are required to operate in different modes with multistandard support. The use of WCDMA (Wideband Code Division Multiple Access) in 3G to achieve high data rates demands higher channel bandwidth. Thus the analogue frontend must be redesigned to handle wider bandwidth with efficient power consumption. To avoid such redesign, it is desirable to have an analogue frontend with programmable bandwidth for multistandard support.

4.6.2   Reconfigurable transceivers using FPAA

The ultimate aim is to produce a single-chip transceiver, which can be dynamically configurable to support many air interfaces, quality of service (QoS) requirements and applications. As discussed above the best way to achieve this is to use reconfigurable hardware, both in the analogue and digital domains of the transceiver. Until now the flexibility provided by reconfigurable hardware has only materialised in digital domain. There have been attempts to move the ADC into the RF stage of the receiver to leave only the anti-aliasing filter in the analogue domain. In order to achieve this, the ADC must digitise the complete incoming RF signal bandwidth. Coping with a signal of such large bandwidth and high dynamic range is a difficult task for ADCs. So this type of ideal software radio may not be possible in practice in the foreseeable future. Positioning the ADC after the down conversion to the IF frequency of the receiver, but before the IF processing, can solve the problem. Here the IF signals

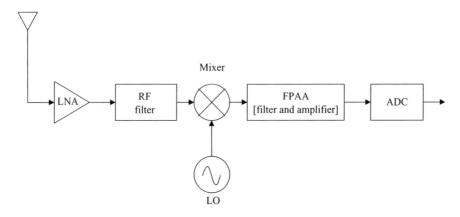

Figure 4.17   Radio receiver with analogue frontend

can be processed digitally in the digital frontend (FPGA). The benefit of using a digital frontend is the very good flexibility because of DSP and increase in system versatility. On the other hand, it also increases power consumption, and high sample rates typically require the use of complex time-shared hardware and increased ADC dynamic range. There are many inherent benefits in implementing the IF stages of the receiver using analogue circuits and techniques, such as smaller silicon area and reduced power consumption, but in doing this we remove the flexibility required by the multistandard receiver. Using reconfigurable analogue devices we could solve this problem.

The radio receiver design using the FPAA to improve the IF stage performance and reconfigurability is shown in Figure 4.17. There are three possible IF structures: high IF, low IF and zero IF. For zero IF, FPAAs using switched-capacitor and active-RC techniques can be used and with current technology, low-IF stages can also be implemented using FPAAs such as those based on the MOSFET-C and OTA-C technique. For a high-IF stage, the OTA-C FPAA may be the only candidate. Note that the zero-IF stage requires lowpass filters and the high-IF stage uses bandpass filters. The low-IF stage may use either lowpass or bandpass filters. From the low power and small die size viewpoints, single amplifier filter structures have been used in some applications [64, 65]. On-chip tuning of high-Q, high-frequency bandpass filters is a big challenge for high-IF design and such tuning circuits must be integrated into the FPAA [39].

Fully reconfigurable receivers may also use an FPAA for the RF frontend. With this fully reconfigurable receiver, software and hardware reconfigurability can be achieved in the RF, IF, as well as baseband. But current FPAAs are still far away from meeting RF requirements.

Finally, the concept of the FPMA may be used, which combines the FPAA (analogue frontend), ADC (interfacing) and FPGA (digital frontend) together, as shown in Figure 4.18.

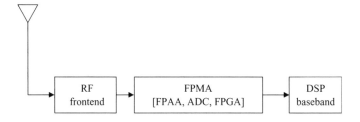

Figure 4.18   Radio receiver using FPMA

Single-chip transceivers require the use of CMOS technology, but the design of analogue and RF circuits in digital CMOS technology entails many difficulties that challenge researchers. Further to transceivers on a single chip, the future market also requires a transceiver on a programmable chip, which will almost certainly see the application of the FPAA and FPMA. This will present a fascinating challenge to FPAA researchers and designers.

4.7   Conclusions

Several FPAA designs have been published in the past few years. Research in this field is actively going on with great interest from the whole analogue sector and will be a key thrust for future analogue design. This chapter has presented an overview of FPAAs. General aspects of FPAAs have been introduced, which include architectures, CABs, CAD, circuit techniques and applications. Commercial FPAAs were then compared and three products illustrated. Some details of the design and implementation of high-frequency continuous-time OTA-C FPAAs have been presented. We have also mentioned FPMAs and systems on a programmable chip. The potential application of FPAAs and FPMAs in future universal wireless communication systems has been particularly discussed.

Although much research has been conducted and some commercial products have been developed, the work is still in its early stage and much remains to be done. The FPAAs published in the literature or which are commercially available generally have low operating frequency and high power consumption. Some also have limited functions and need external components. FPAAs have been used for quick analogue circuit design in many areas including teaching in higher education institutions. However, there have not been wide applications in communication and information systems.

Lower supply voltage and power consumption, higher frequency operation and larger dynamic range are required for many applications such as wireless communication and computer hard disk systems. Consumer electronic systems also often require very high performance analogue circuits. Low-power and high-frequency CMOS FPAAs will continue to be an active area of development. General-purpose FPAAs will be further developed and the issues of applying application-specific FPAAs embedded in fully reconfigurable electronic systems will have some particular interest. The design of universal transceivers for multistandard mobile communications (GSM, UMTS, etc.) and short-range wireless communications (Bluetooth, WLAN, etc.) using FPAAs will be an attractive topic.

In the past, active simulation of the inductor was utilised due to the difficulty in integrating the passive inductor on silicon. However, active inductor simulation cannot meet the requirements of RF wireless communications and bulky discrete inductors have had to be used in many RF applications. Attempts to put inductors on silicon have thus been more actively pursued recently due to the exploding wireless market [66]. As FPAA's operation frequencies move to RF, CMOS silicon inductors may need to be considered as an elementary component for the CABs. MEMS techniques have also been used to integrate inductors on chips.

Industry has been in the race to make reconfigurable analogue devices for some years and they are now trying to make a reconfigurable mixed-signal device. The latter proves even more difficult and very much effort is needed to make a FPMA with reasonable performance. Implementation of FPMAs is important for SoC and SoPC designs and will be an active subject of future research.

4.8   References

1 SIVILOTTI, M. A.: 'A dynamically configurable architecture for prototyping analog circuits'. Proceedings of Decennial Caltech Conference on *Advanced Research in VLSI*, Cambridge, MA, March 1988, pp. 237–58

2 GAUDET, V. C. and GULAK, P. G.: 'Implementation issues for high-bandwidth field-programmable analog arrays'. *Journal of Circuits, Systems and Computers*, 2000, **8** (5), pp. 541–58

3 LOOBY, C. A. and LYDEN, C.: 'Op-amp based CMOS field-programmable analogue array'. IEE Proceedings: *Circuits, Devices and Systems*, April 2000, **147** (2), pp. 93–9

4 LEE, E. K. F. and GULAK, P. G.: 'A CMOS field programmable analog array', *IEEE Journal of Solid-State Circuits*, December 1991, **26** (12), pp. 1860–67

5 LEE, E. K. and GULAK, P. G.: 'A transconductor-based field-programmable analog array'. IEEE International Solid-State Circuits Conference Digest, San Francisco, CA, USA, February 1995, pp. 198–99

6 EMBABI, S. H. K., QUAN, X., OKI, N., MANJREKAR, A. and SANCHEZ-SINENCIO, E.: 'A field programmable analog signal processing array'. Proceedings of IEEE Midwest Symposium on *Circuits and Systems*, Puebla, August 1996, pp. 151–54

7 QUAN, X., EMBABI, S. H. K. and SANCHEZ-SINENCIO, E.: 'A current-mode based field programmable analog array architecture for signal processing applications'. Proceedings of IEEE Custom Integrated Circuits Conference, Santa Clara, 1998, pp. 12.6.1–12.6.4

8 DEAKIN, S. J. and HATFIELD, J.: 'A field programmable current-mode analog VLSI'. Proceedings of IEE Symposium on *Analogue Signal Processing*, Oxford, November 1996, pp. 6/1–6/6

9 PIERZCHALA, E. and PERKOWSKI, M. A.: 'High-speed field programmable analog array architecture design'. Proceedings of ACM/SIGDA FPGA'94, Berkeley, CA, USA, February 1994

10 PIERZCHALA, E., PERKOWSKI, M., VAN HALEN, P. and SCHAUMANN, R.: 'Current-mode amplifier/integrator for a field programmable analog array'. IEEE International Solid-State Circuits Conference Technical Digest, San Francisco, February 1995, pp. 196–97

11 KUTUK, H. and KANG, S. M.: 'A field-programmable analog array (FPAA) using switched-capacitor technique'. Proceedings of IEEE International Symposium on *Circuits and Systems*, Atlanta, May 1996, **4**, pp. 41–43

12 KLEIN, H. W.: 'The EPAC architecture: an expert cell approach to field programmable analog devices'. Proceedings of ACM/SIGDA FPGA'96, Monterey, CA, 1996, pp. 94–8

13 BRATT, A. and MACBETH, I.: 'Design and implementation of a field programmable analogue array'. Proceedings of ACM/SIGDA FPGA'96, Monterey, CA, 1996, pp. 88–93

14 ANDERSON, D., MARCJAN, C., BERSCH, D., ANDERSON, H., HU, P., PALUSINSKI, O., GETTMAN, D., MACBETH, I. and BRATT, A.: 'A field programmable analog array and its application'. IEEE Custom Integrated Circuits Conference, Santa Clara, 1997, pp. 555–58

15 GAUDET, V. C. and GULAK, P. G.: 'CMOS implementation of a current conveyor-based field-programmable analog array' (University of Toronto, 1998), pp. 1156–59

16 CHANG, S. T., HAYES-GILL, B. R. and PAUL, C. J.: 'Multifunction block for a switched-current field programmable analogue array'. Proceedings of IEEE Midwest Symposium on *Circuits and Systems*, Ames, Iowa, August 1996, **1**, pp. 8–16

17 BOZIC, M. and GRUNDY, D. L.: 'A novel approach to VLSI analogue design'. Proceedings of IEE Symposium on *Analogue Signal Processing*, Oxford, 1996, pp. 10/1–10/6

18 PANKIEWICZ, B., WÓJCIKOWSKI, M., SZCZEPANSKI, S. and SUN, Y.: 'A CMOS field programmable analog array and its application in continuous-time OTA-C filter design'. Proceedings of IEEE International Symposium on *Circuits and Systems*, Sydney, May 2001, **1**, pp. 5–8

19 PANKIEWICZ, B., WÓJCIKOWSKI, M., SZCZEPAÑSKI, S. and SUN, Y.: 'A CMOS field programmable analog array for continuous-time OTA-C filter applications', *IEEE Journal of Solid-State Circuits*, February 2002, **147** (2), pp. 125–36

20 LOH, K. H., HISER, D. L., ADAMS, W. J. and GEIGER, R. L.: 'A versatile digitally controlled continuous-time filter structure with wide range and fine resolution capability', *IEEE Transactions on Circuits and Systems-II*, 1992, **39**, pp. 265–76

21 MOTOROLA, 'MPAA020 field programmable analog array datasheet'. April 1997

22 IMP Inc, 'Preliminary Product Information – IMP50E10 EPAC (Electrically programmable analog circuit)'. Datasheet, 1995; EPAC Design Handbook, April 1996

23 Lattice Semiconductor, Programmable analogue circuits (ispPAC). [Online] http://www.latticesemi.com

24 Zetex Semiconductors, TRAC (Totally Reconfigurable Analogue Circuit). [Online]. http://www.zetex.com

25 Anadigm Company, The AN10E40 field programmable analog array. [Online]. http://www.anadigm.com

26 SIDSA Company, FIPSOC mixed signal system on chip. [Online]. http://www.fipsoc.com

27 Altera Company, SoPC Information [Online], http://www.altera.com

28 TOUMAZOU, C., LIDGEY, F. J. and HAIGH, D. G. (Eds.): 'Analogue IC design: the current-mode approach' (Peter Peregrinus, London, 1990)

29 LAKER, K. R. and SANSEN, W.: 'Design of analog integrated circuits and systems' (McGraw-Hill, 1994)

30 JOHNS, D. A. and MARTIN, K.: 'Analog integrated circuit design' (John Wiley, New York, 1997)

31 SANCHEZ-SINENCIO, E. and ANDREOU, A. G. (Eds.): 'Low-voltage/low-power integrated circuits and systems' (IEEE Press, Piscataway, New Jersey, 1999)

32 TSIVIDIS, Y. and VOORMAN, J. O. (Eds.): 'Integrated continuous-time filters' (IEEE Press, Piscataway, New Jersey, 1993)

33 SCHAUMANN, R., LAKER, K. R. and GHAUSI, M. S.: 'Active filter design: passive, active and switched-capacitor' (Prentice Hall, Englewood Cliffs, New Jersey, 1990)

34 DELIYANNIS, T., SUN, Y. and FIDLER, J. K.: 'Continuous-time active filter design' (CRC Press, Florida, USA, January 1999)

35 SUN, Y.: 'Design of high frequency integrated analogue filters' (IEE, Stevenage, UK, 2002)

36 SUN, Y.: Guest Editor, 'Special issue on high-frequency integrated analogue filters', *IEE Proceedings: Circuits, Devices and Systems*, February 2000, **147** (1), pp. 1–90

37 SCHAUMANN, R. and KARSILAYAN, A. I.: 'On-chip automatic tuning of filters', Chapter 7, in Sun, Y. (Ed.): 'Design of high frequency integrated analogue filters' (IEE, Stevenage UK, 2002)

38 MORITZ, J. R. and SUN, Y.: 'Tuning of multiple loop feedback leapfrog bandpass filters', *Electronics Letters*, 2001, **37** (11), pp. 671–72

39 MORITZ, J. R. and SUN, Y.: 'Automatic tuning of high-frequency high-Q multiple loop feedback bandpass filters'. Proceedings of IEEE International Symposium on *Circuits and Systems*, Arizona, USA, May 2002

40 HUGHES, J. B., WORAPISHET, A. and SITDHIKORN, R.: 'Low voltage techniques for switched-current filters', Chapter 5, in Sun, Y. (Ed.): 'Design of high frequency integrated analogue filters' (IEE, Stevenage, UK, 2002)

41 FREY, D.: 'Log Domain Filters', Chapter 4, in Sun, Y. (Ed.): 'Design of high frequency integrated analogue filters' (IEE, Stevenage, UK, 2002)

42 BANU, M. and TSIVIDIS, Y.: 'MOSFET-C techniques: designing power efficient high frequency filters', Chapter 2, in Sun, Y. (Ed.): 'Design of high frequency integrated analogue filters' (IEE, Stevenage, UK, 2002)

43 GROENEVOLD, G.: 'Low-power MOSFET-C 120MHz Bessel allpass filter with extended tuning range', *IEE Proceedings: Circuits, Devices and Systems*, 2000, **147** (1), pp. 28–34

44 HE, Y., JIANG, J. and SUN, Y.: 'CMOS R-MOSFET-C fourth-order Bessel filter with accurate group delay', Proceedings of IEEE International Symposium on *Circuits and Systems*, Arizona, USA, May 2002

45 SUN, Y. and HILL, C.: 'Low power fully differential CMOS filter for video frequencies', *IEEE Transactions on Circuits and Systems, Part-II: Analog and Digital Signal Processing*, 2001, **48** (12), pp. 1144–48

46 SANCHEZ-SINENCIO, E. and SILVA-MARTINEZ, J.: 'CMOS transconductance amplifiers, architectures and active filters: a tutorial', *IEE Proceedings: Circuits, Devices and Systems*, **147** (1), 2000, pp. 3–12

47 GLINIANOWICZ, J., JAKUSZ, J., SZCZEPANSKI, S. and SUN, Y.: 'A high-frequency two-input CMOS OTA for continuous-time filter applications', *IEE Proceedings: Circuits, Devices and Systems*, 2000, **147** (1), pp. 13–18

48 SUN, Y., HILL, C. and SZCZEPANSKI, S.: 'Large dynamic range high frequency fully differential CMOS transconductance amplifier', *International Journal of Analog Integrated Circuits and Signal Processing*, March 2003, **34** (3), pp. 247–55

49 SUN, Y.: 'Architectures and design of OTA/gm-C filters', Chapter 1, in Sun, Y. (Ed.): 'Design of high frequency integrated analogue filters' (IEE, Stevenage, UK, 2002)

50 SANCHEZ-SINENCIO, E. and SMITH, S. L.: 'Continuous-time low-voltage current-mode filters', Chapter 11, in Sanchez-Sinencio, E. and Andreou, A. G. (Eds.): 'Low-voltage/low-power integrated circuits and systems' (IEEE Press, Piscataway, New Jersey, 1999)

51 HAN, G. and SANCHEZ-SINENCIO, E.: 'CMOS transconductance multipliers: a tutorial', *IEEE Transactions on Circuits and Systems II: Analog and Digital Signal Processing*, 1998, **45**, pp. 1550–63

52 PANKIEWICZ, B., SZCZEPANSKI, S. and SUN, Y.: 'CMOS level shifter and four quadrant analog mulitplier'. Proceedings of 9th International Conference on *Mixed Design of Integrated Circuits and Systems*, Wroclaw, Poland, June 2002

53 LINARES-BARRANCO, B., RODRIGUEZ-VAZQUEZ, A., SANCHEZ-SINENCIO, E. and HUERTAS, J. L.: 'Generation, design and tuning of OTA-C high-frequency sinusoidal oscillators', *IEE Proceedings-G*, 1992, **139** (5), pp. 557–68

54 SANCHEZ-SINENCIO, E., RAMIREZ-ANGULO, J., LINARES-BARRANCO, B. and RODRIGUEZ-VAZQUEZ, A.: 'Operational transconductance amplifier-based nonlinear function synthesis', *IEEE Journal of Solid-State Circuits*, 1989, **24** (6), pp. 1576–85

55 LINARES-BARRANCO, B., SANCHEZ-SINENCIO, E., RODRIGUEZ-VAZQUEZ, A. and HUERTAS, J. L.: 'A modular T-mode design approach for analog neural network hardware implementations', *IEEE Journal of Solid-State Circuits*, 1992, **27** (5), pp. 701–13

56 THAPAR, H., LEE, S. S., CONROY, C., CONTRERAS, R., YEUNG, A., CHERN, J.G., PAN, T. and SHIH, S. M.: 'Hard disk drive read channels: technology and trends'. Proceedings of IEEE Custom Integrated Circuits Conference, Santa Clara, 1998, pp. 309–16

57 SU, H. and SUN, Y.: 'A CMOS 100MHz continuous-time seventh-order 0.05 equiripple linear phase leapfrog multiple loop feedback gm-C filters'. Proceedings of IEEE International Symposium on *Circuits and Systems*, Arizona, May 2002

58 RAZAVI, B.: 'Challenges in portable RF transceiver design', *IEEE Circuits and Devices Magazine*, September 1996, pp. 12–25

59 ABIDI, A.: 'Low-power radio-frequency IC's for portable communications', *Proceedings of the IEEE*, April 1995, **83** (4), pp. 544–69,

60 SUN, Y.: Guest Editor, 'Special Issue on RF Circuits and Systems for Wireless Communications', *IEE Proceedings: Circuits, Devices and Systems*, 2000, **149** (5), pp. 321–75

61 COLSELL, S.: and EDWARDS, R.: 'A comparative study of reconfigurable digital and analogue technologies for future mobile communication systems'. Proceedings of IEE International Conference 3G Mobile Communication Technologies, London, March 2001, pp. 302–5

62 HOLLAMAN, T., LINDFORS, S., SALO, T., LANSIRINNE, M. and HALONEN, K.: 'A 2.7 CMOS dual-mode basedband filter for GSM and WCDMA'. Proceedings of IEEE ISCAS, Sydney, Australia, 2001, pp. I-316–9

63 ALZAHER, H. A., ELWAN, H. O. and ISMAIL, M.: 'A CMOS highly linear channel-select filter for 3G multistandard integrated wireless receivers', *IEEE Journal of Solid-State Circuits*, 2002, **37** (1), pp. 27–37

64 ELWAN, H., ALZAHER, H. and ISMAIL, M.: 'A new generation of global wireless compatibility', *IEEE Circuits and Devices Magazine*, January, 2001, pp. 2–19

65 ELWAN, H., RAVINDRAN, A. and ISMAIL, M.: 'A CMOS low power baseband chain for a GSM/DECT multistandard receiver', *IEE Proceedings: Circuits, Devices and Systems*, 2002, **149** (5), pp. 337–47

66 LI, D. and TSIVIDIS, Y.: 'Active filters using integrated inductors', Chapter 3, in Sun, Y. (Ed.): 'Design of high frequency integrated analogue filters' (IEE, Stevenage, UK, 2002)

Chapter 5

A low-power, low-voltage Bluetooth channel filter using class AB CMOS transconductors

John Hughes, Adrian Spencer, Apisak Worapishet and Rungsimant Sitdhikorn

5.1 Introduction

The low-IF (intermediate frequency) polyphase architecture has emerged as the preferred approach for achieving the required sensitivity in fully integrated wireless transceivers [1]. The industry is currently driving down the cost and power consumption by attempting standard CMOS solutions for applications such as Bluetooth [2] and ZigBee (HomeRF Lite) [3] and this is creating new challenges for the circuit designer.

Figure 5.1 shows a typical low-IF receiver architecture. One of the keys to the success of this polyphase approach has been the ability to integrate the channel filter. However, when solutions are required using the ever lowering supply voltages demanded by standard CMOS, and with perhaps an order of magnitude lower

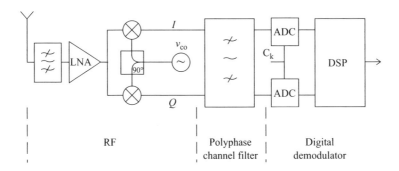

Figure 5.1   Typical low-IF receiver architecture

power consumption, conventional filter techniques using operational amplifiers with passive resistors and capacitors may prove inadequate.

The approach adopted for this work was to make transconductor-capacitor (Gm-C) filters using class AB CMOS transconductors. While these can be expected to yield more efficient solutions than the conventional approach, they would not be expected to be as linear as the alternative passive resistors. So, a key aspect of this work was to demonstrate that suitable filter responses could be achieved with considerably lower power consumption and with adequate intermodulation performance.

In this chapter, we design a Gm-C complex low-IF channel filter for Bluetooth. The class AB transconductor is developed and simulated performance of the filter designed with these transconductors is presented.

5.2   Filter synthesis

The function of the channel filter is to pass the wanted channel while rejecting neighbouring channel interferers. As these interferers occur at frequencies on either side of the passband, image responses must be suppressed and this requires using a bandpass filter with an asymmetric amplitude response, i.e. $|H(j\omega)| \neq |H(-j\omega)|$. Filters of this type require complex coefficients and these can be created by polyphase networks employing two paths driven by signals which are identical but in exact phase quadrature as supplied in the down-converted I- and Q-channels of a low-IF transceiver.

5.2.1   Complex filters

The principle of the complex filter is illustrated in Figure 5.2. Starting with a real lowpass filter (i.e. with real coefficients), the transformation $s \rightarrow s - j\omega_0$ is applied. This shifts the poles up the imaginary axis by ω_0 and transforms the lowpass response (actually a bandpass response centred at $\omega = 0$) into an identical bandpass response centred at $\omega = \omega_0$ [4]. The transformation preserves both

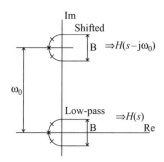

Figure 5.2   Complex filter basics

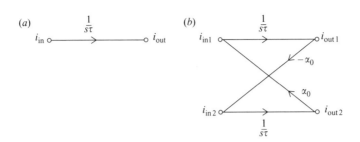

Figure 5.3 Ideal integrator signal flow-graphs: (a) real, (b) complex

amplitude and phase characteristics and produces the required feature of having no image response at negative frequency.

Synthesis of complex filters follows similar procedures to those for real filters except that it makes use of complex integrators. Figure 5.3 shows the signal flow graphs of real and complex current-mode integrators (voltage-mode integrators will have identical signal flow graphs).

The real integrator has an input i_{in} and an output i_{out}, and a transfer characteristic described by

$$H(s) = \frac{i_{out}(s)}{i_{in}(s)} = \frac{1}{s\tau} \tag{5.1}$$

where τ is the integrator time constant.

The complex integrator has two inputs i_{in1} and i_{in2} where $i_{in2} = j \cdot i_{in1}$, two outputs i_{out1} and i_{out2} and two extra outputs $-\alpha_0 \cdot i_{out1}$ and $\alpha_0 \cdot i_{out2}$ fed back to the inputs where they sum with i_{in2} and i_{in1} respectively. We can demonstrate that this is the required signal flow graph from its nodal equations which are

$$i_{out1}(s) = (i_{in1}(s) + \alpha_0 i_{out2}(s))\frac{1}{s\tau} \tag{5.2}$$

$$i_{out2}(s) = (i_{in2}(s) - \alpha_0 i_{out1}(s))\frac{1}{s\tau} \tag{5.3}$$

Combining these equations, the transfer characteristics for the two paths are

$$H_1(s) = \frac{i_{out1}(s)}{i_{in1}(s)} = \frac{1}{s\tau - j\alpha_0} \tag{5.4}$$

$$H_2(s) = \frac{i_{out2}(s)}{i_{in2}(s)} = \frac{1}{s\tau - j\alpha_0} \tag{5.5}$$

So, $H_1(s) = H_2(s) = H(s)$ and the outputs $i_{out1}(s)$ and $i_{out2}(s)$ are identical but in phase quadrature. For $\alpha_0 = \omega_0\tau$,

$$H(s) = \frac{1}{(s - j\omega_0)\tau} \tag{5.6}$$

which demonstrates that the transformation, $s \rightarrow s - \omega_0 \tau$, is being performed as required by the complex integrator.

The physical implementation of a complex filter is determined by the chosen integrator realisation. Traditionally, continuous-time RC-active integrators [5] using passive resistors and capacitors, and operational amplifiers or special buffers [6], are employed. However, it will be difficult to maintain their excellent performance as we use lower supply voltage CMOS processes. Sampled-analogue integrators using switched capacitors [7] or switched currents [8] may also be employed. Switched capacitors will also lose performance in low-voltage CMOS and, being sampled-analogue, require additional anti-alias filters. Switched currents [8] also require additional anti-alias filters but their performance is not expected to degrade at lower supply voltages [9].

Here the plan is to use the Gm-C technique and Figure 5.4 shows current-mode integrators for use in our Gm-C filters. The real integrator input current flows into the capacitor C to develop an input voltage $v_{\text{in}}(s) = i_{\text{in}}(s)/sC$ which produces an output current $i_{\text{out}}(s) = G \cdot v_{\text{in}}(s) = i_{\text{in}}(s) \cdot G/sC$. The transfer function is $H(s) = G/sC$ which describes a real integrator with a time constant $\tau = C/G$. The complex integrator is developed directly from its signal flow graph and the feedback signals are produced by the cross-branch transconductors G_0 and $-G_0$ where $G_0 = \alpha_0 \cdot G$. The frequency shift ω_0 is given by

$$\omega_0 = \frac{\alpha_0}{\tau} = \frac{G_0}{C} \tag{5.7}$$

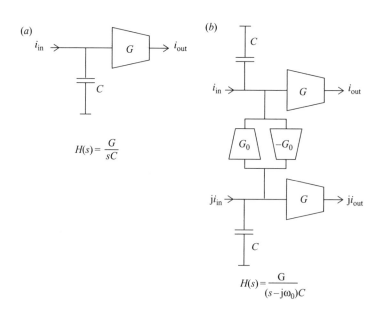

$$H(s) = \frac{G}{sC}$$

$$H(s) = \frac{G}{(s - j\omega_0)C}$$

Figure 5.4   *Current-mode Gm-C integrators: (a) real, (b) complex*

5.2.2   Bluetooth filter synthesis

We can now proceed to synthesise the complex filter. First, we generate the lowpass real filter by traditional leapfrog methods and then the complex bandpass filter is created by shifting the real lowpass response to the required centre frequency. We will create a fifth-order 0.5 dB equiripple Chebyshev complex leapfrog channel filter suitable for Bluetooth, using the Gm-C current-mode integrators described in the previous section.

The synthesis starts with the lowpass *LCR* prototype shown in Figure 5.5a. Its nodal equations are:

$$v_1 = \frac{1}{sC_1}\left(i_{in} - \frac{v_1}{R} - i_2\right) \tag{5.8}$$

$$i_2 = \frac{1}{sL_2}(v_1 - v_3) \tag{5.9}$$

$$v_3 = \frac{1}{sC_3}(i_2 - i_4) \tag{5.10}$$

$$i_4 = \frac{1}{sL_4}(v_3 - v_{out}) \tag{5.11}$$

$$v_{out} = \frac{1}{sC_5}\left(i_4 - \frac{v_{out}}{R}\right) \tag{5.12}$$

Using traditional signal flow graph methods, the Gm-C filter structure shown in Figure 5.5b is constructed which simulates the nodal equations of the prototype using state variables v_1, i_2, v_3, i_4 and v_5. The specific current integrators are shown boxed.

Translating this lowpass design into its complex bandpass counterpart involves providing two paths, each containing the lowpass filter, and then replacing the real integrators with complex integrators. This results in the architecture shown in Figure 5.5c. The design values for the Bluetooth channel filter response ($F_0 = 1\,\text{MHz}$, $\text{BW} = 1.2\,\text{MHz}$) are given in Table 5.1.

It can be seen from Figure 5.5c that the various transconductor loops are in fact gyrators [10]. Figure 5.6 shows a gyrator view of the channel filter. The circuit clusters at C_2 and C_4 are gyrator-C equivalents for the inductors L_2 and L_4. Furthermore, the cross-branch transconductor loops which produce the frequency shift are also gyrators which produce fixed negative susceptances of $-j\omega_0 C_n$ at each capacitor C_n. The capacitor set is transformed from C_n to $(s - j\omega_0)C_n$ which again explains the frequency shift.

A balanced version of the channel filter is shown in Figure 5.7 and the design values of its differential transconductors and floating capacitors are half of those of the single-ended design given in Table 5.1. Using idealised transconductors and capacitors, the simulated idealised amplitude response is shown in Figure 5.8 and the idealised passband amplitude and group delay responses are shown in Figure 5.9.

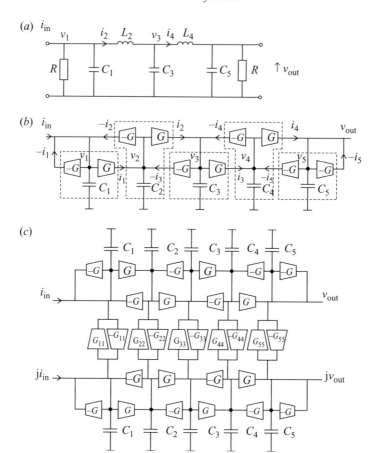

Figure 5.5   *Channel filter architecture: (a) fifth-order LP prototype; (b) Gm-C LP filter; (c) Gm-C complex bandpass filter*

5.3   Class AB CMOS transconductor

Ideally, a transconductor linearly converts an input voltage into an output current with both input and output ports presenting infinite impedance. Our transconductor must be implemented in standard digital CMOS and so we must find a solution which has the required high performance while satisfying the voltage headroom constraints imposed by the process.

5.3.1   Cell development

One promising approach which uses the PMOS/NMOS transistor pair proposed by Nauta [11] for mainly high-frequency filters and employed recently by Andreani [12] at lower frequencies is shown in Figure 5.10a. If the PMOS and NMOS transistors

Table 5.1 *Single-ended design values of Bluetooth channel filter*

Parameter	Value
G	$40\,\mu\text{S}$
G_{11}	$113.72\,\mu\text{S}$
G_{22}	$81.97\,\mu\text{S}$
G_{33}	$169.39\,\mu\text{S}$
G_{44}	$81.97\,\mu\text{S}$
G_{55}	$113.72\,\mu\text{S}$
C_1	$18.10\,\text{pF}$
C_2	$13.05\,\text{pF}$
C_3	$26.96\,\text{pF}$
C_4	$13.05\,\text{pF}$
C_5	$18.10\,\text{pF}$

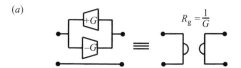

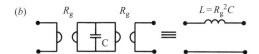

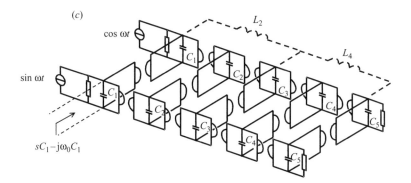

Figure 5.6 *Gyrator view of channel filter: (a) gyrator–transconductor equivalence; (b) gyrator-C inductor; (c) channel filter*

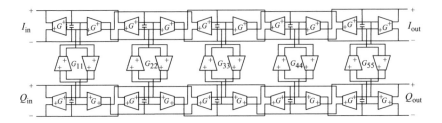

Figure 5.7   Balanced channel filter

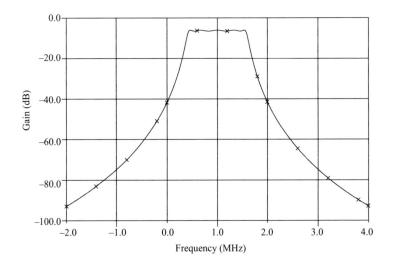

Figure 5.8   Ideal amplitude response of balanced channel filter

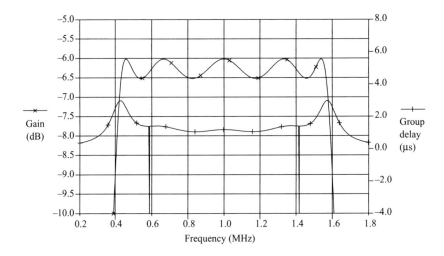

Figure 5.9   Ideal amplitude and group delay responses of balanced channel filter

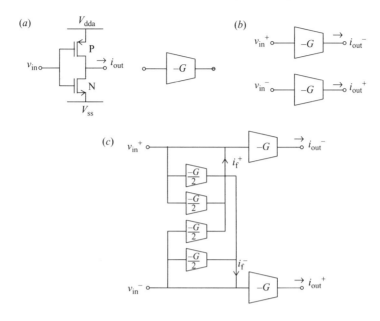

Figure 5.10   Class AB CMOS transconductors: (a) simple single-ended; (b) simple balanced; (c) fully differential

P and N have identical parameters (this is an unnecessary constraint but one which simplifies the description), then $g_{mp} = g_{mn} = g_m$ and the overall transconductance of this single-ended cell is $-G$ where $G = 2 \cdot g_m$. Biasing the input at the mid-rail voltage ($V_{dda}/2$) produces equal drain currents J in both P and N and the output current is zero. When the input voltage changes by v_{in}, the drain currents of N and P are unbalanced as given by

$$I_{n,p} = J \left(1 \pm \frac{v_{in}G}{4J} \right)^2 \tag{5.13}$$

and a linearly related current, $i_{out} = I_p - I_n = -Gv_{in}$, flows at the output. This theoretical linearity occurs despite the square law relationships determining the individual currents as shown in normalised form in Figure 5.11. The transconductor is exceptionally efficient because it operates in class AB for peak output currents as high as $i_{out} = 4J$.

One feature of this transconductor is that G is influenced by the bias current J which is strongly determined by the supply voltage, V_{dda}. This feature allows either the value of G or the frequency tuning of the filter to be simply controlled using control circuits like those described later. It also carries a penalty for power supply noise feedthrough. Assuming square-law saturated MOS behaviour, the transconductance

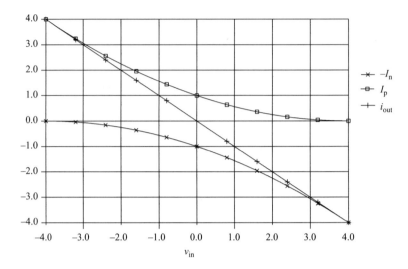

Figure 5.11   Normalised (J = 1, G = 1) *behaviour of the class AB transconductor*

is given by

$$G = g_{mp} + g_{mn} = 2kV_{gt} = 2k\left(\frac{V_{dda}}{2} - V_t\right) \tag{5.14}$$

where $k = 0.5\,\mu C_{ox} W/L$. If now the value of V_{dda} is modulated by a signal v_d then G is also modulated

$$G(v_d) = 2k\left(\frac{V_{dda} + v_d}{2} - V_t\right) \tag{5.15}$$

and the output current is given by

$$i_{out} = G(v_d)v_{in} \tag{5.16}$$

$$= Gv_{in} + kv_{in}v_d \tag{5.17}$$

So, in addition to the wanted output signal, $-Gv_{in}$, there is an extraneous signal, $kv_{in}v_d$, which results in intermodulation of signals with supply noise. Consequently, great care must be taken when designing the V_{dda} control circuit so that both V_{dd} and V_{ss} noise is effectively suppressed.

Balanced transconductors (which are able to convert a balanced input voltage into a balanced output current) are highly desirable. In Gm-C filters, inversion of signals is frequently required and this can be achieved by simply crossing over signal pairs. A simple balanced arrangement may be formed from two of the single-ended transconductor cells as shown in Figure 5.10b, one converting the positive signal voltage and the other converting the negative signal voltage, and achieves a differential transconductance $G_m = G/2$. Unfortunately, if this simple arrangement is used in feedback networks, such as the gyrator loops occurring in a Gm-C filter, the circuits become unstable.

This can be resolved with a fully differential transconductor such as that shown in Figure 5.10c [11]. It comprises two main single-ended transconductors $(-G)$ and a common-mode feedback network between the input ports. This employs four half-size single-ended transconductors $(-G/2$, each using half-width transistors and half the bias current) coupled between the inputs of the main transconductors.

First, consider this transconductor under quiescent conditions, i.e. $v_{in}^+ = v_{in}^- = 0$, i.e. both inputs are at $V_{dda}/2$ and the output currents are zero. The current in each of the common-mode feedback MOSTs is $J/2$ and the feedback currents are $i_f^+ = i_f^- = 0$.

Next, consider the transconductor with purely differential input voltages, i.e. $v_{in}^+ = v_{dm}/2$ and $v_{in}^- = -v_{dm}/2$ where v_{dm} is the differential mode input voltage. The feedback currents are again $i_f^+ = i_f^- = 0$ because the half-size transconductor pairs generate equal and opposite currents that cancel. The voltages v_{in}^+ and v_{in}^- are applied directly to the main transconductor inputs and a current of $-v_{dm} \cdot G/2$ flows at the outputs.

Finally, consider the transconductor with common-mode input currents, i_{cm}. These force the input voltages, v_{in}^+ and v_{in}^-, until $i_f^+ = i_f^- = i_{cm}$. This occurs when $v_{in}^+ = v_{in}^- = i_{cm}/G$ and this causes a common–mode output current $i_{out}^+ = i_{out}^- = -i_{cm}$. So, while the transconductor has not rejected the common-mode input it is nevertheless stable.

One problem with this transconductor arises because of feedthrough from input to output via the gate–drain capacitances. Had these capacitances been reciprocal (i.e. $C_{gd} = C_{dg}$) then the balanced gyrators formed by the $-G$, $+G$ loops in a Gm-C filter (see Figure 5.12) would have equal feedforward and feedback capacitances

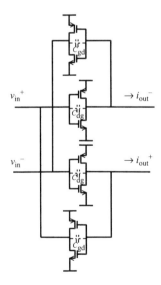

Figure 5.12   Gyrator loop of two transconductors with feedthrough capacitance

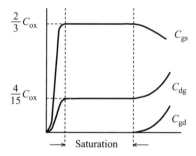

Figure 5.13   Capacitances of a MOS transistor in saturation

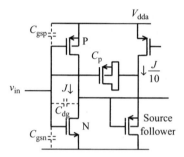

Figure 5.14   Transconductor with capacitive feedthrough equalisation

which produce no resultant feedthrough. Unfortunately, the capacitances of a MOS transistor in saturation are non-reciprocal [13] as shown in Figure 5.13. As a saturated MOS transistor is 'pinched-off' at its drain, voltage disturbances at the drain do not influence the channel charge and so $C_{gd} \approx 0$. On the other hand, voltage disturbances at the gate directly influence the channel charge (to produce a change in the drain current) and so $C_{dg} \approx \frac{4}{15} C_{ox} \approx \frac{2}{5} C_{gs}$. In a gyrator loop, the strong feedforward via C_{dg} in transconductor $- G$ (or $+ G$) is not balanced by an equal feedback via C_{gd} in transconductor $+ G$ (or $- G$) and this can produce filter responses which peak at high frequency.

The solution to this is shown in Figure 5.14. The pMOS capacitors C_p are connected between each input and a source follower connected to each output. Voltage disturbances at the transconductor input produce feedforward via C_p but the capacitive current is routed harmlessly via the source follower to V_{ss} and the transconductor experiences no extra feedforward. However, voltage disturbances at the transconductor outputs do produce feedback via C_p. Clearly, C_p creates only capacitive *feedback* while the internal transconductors produce only *feedforward*. If we make $C_p = C_{dgp} + C_{dgn} \approx \frac{2}{5}(C_{gsp} + C_{gsn})$, then the fully differential transconductor has reciprocal feedthrough capacitance. Now when the transconductors are connected as gyrators, the feedforward cancels the feedback and feedthrough is eliminated.

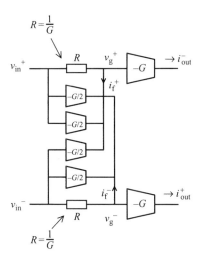

Figure 5.15   Transconductor with improved common-mode rejection

A further enhancement to the transconductor of Figure 5.10c which improves its common-mode rejection ratio is shown in Figure 5.15. The extra resistor R in the common-mode feedback network is given the value $R = 1/G$. With purely common-mode input currents, i_{cm}, the input voltages, v_{in}^+ and v_{in}^-, adjust until $i_f^+ = i_f^- = i_{cm}$ as before. This occurs when $v_{in}^+ = v_{in}^- = i_{cm}/G$ but now i_{cm} flows via resistor R to produce an extra voltage drop of $-i_{cm}/G$. Consequently, the transconductor input voltage stays at its quiescent value ($V_{dda}/2$) and the common-mode input is rejected. With purely differential signals, $i_f^+ = i_f^- = 0$ as before and there is no voltage drop across R.

So, in a Gm-C filter, we would use the transconductor with feedthrough equalisation (Figure 5.14) in all gyrator loops but we may use the transconductor with improved common-mode rejection ratio (Figure 5.15) at the filter terminations.

5.3.2   Frequency tuning

One feature of the class AB transconductor cells is their sensitivity to the supply voltage. They are only practicable when operated from a regulated supply and this may be integrated with a coarse automatic frequency tuning scheme as shown in Figure 5.16a. Greater precision would be possible with a phase-locked loop [14] but as the required tuning accuracy was only ± 6 per cent the simpler arrangement described here was adopted to save power consumption and chip area. The arrangement uses a control loop comprising a skewed 'diode-connected' transconductor $-G_0$ which generates a quiescent voltage which is offset from that of the reference transconductor, a parallel arrangement of the reference transconductor $-G$ and switched capacitor C, an integrator comprising $-G_{int}$ and C_{int}, a digital inverter and a charge pump Q_{pump} for regulating the V_{dda0} supply. In operation, the reference transconductor and switched

(a)

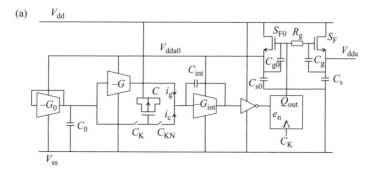

(b)

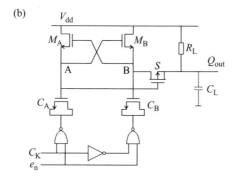

Figure 5.16   Combined tuning and supply regulator arrangement

capacitor supply the integrator with opposite polarity currents, i_g and i_c, which are integrated and amplified and then used to control the charge pump. With the switched capacitor set to the value G/F_{ck} the currents sum to zero and the loop stabilises with the value of V_{dda0} required to achieve the desired time-constant. When the transconductor or capacitor have non-typical values, V_{dda0} adjusts to change the transconductance value to restore the time constant to its nominal value. Necessarily, the loop regulates with a ripple on V_{dda0} and so the supply voltage V_{dda} used by the filter is lowpass filtered by R_g and C_g. With the source follower regulating transistors SF_0 and SF scaled to produce the same V_{gs}, the value of V_{dda} tracks closely that of V_{dda0} and the filter's transconductors will closely track the control loop's reference transconductor. The decoupling capacitors C_{s0}, C_{g0}, C_s and C_g were all nMOS oxide capacitors connected to minimise their gate voltages. They also serve the purpose of transmitting any substrate interference onto the V_{dda} rail. With near equal interference on both V_{ss} and V_{dda} rails very little signal is coupled into the filter.

The charge pump [15] is shown in Figure 5.16b and operates when its enable input, e_n, is high and charge is pumped into the load capacitance via the pMOS switch S to produce a voltage at Q_{out} which can rise above the V_{dd} supply voltage. The resistor R_L acts as a pull-up resistor when Q_{out} is low during start-up and as a pull-down resistor when Q_{out} is high and this allows the loop to stabilise satisfactorily.

The tuning loop was designed for a clock frequency of 26 MHz. The transconductor $-G_0$ produces an offset voltage of about 80 mV. The integrator transconductor $-G_{int}$ was the same as the reference transconductor $-G$ but 20 times larger and so produced the same quiescent input voltage. The capacitors C_0 and C_{int} were each about 20 times larger than the pMOS switched capacitor C, the value of which was adjusted, with typical processing and temperature, until the stabilised value of V_{dda} had the value needed to tune the filter to its nominal response. The cut-off frequency of the lowpass filter formed by R_g and C_g was made sufficiently low to reduce the ripple on V_{dda} to less than 0.1 mV.

5.4   Filter design

The balanced channel filter (Figure 5.7) was implemented in a 2.5 V, 0.25 μm, standard CMOS process. It employed feedthrough-equalised fully differential transconductors operating from a supply voltage (V_{dd}) of 2 V and an internal analogue supply (V_{dda}) of 1.58 V. The I- and Q-channel transconductors all had differential transconductance of 20 μS ($G = 40$ μS) while the cross-branch transconductors were of the same design but with their transistor widths and currents scaled appropriately. This value of transconductance was necessary to achieve the required signal-to-noise ratio. The common-mode feedback networks (not shown for simplicity) were connected across each capacitor and scaled according to the total transconductance that they served. The floating capacitors were made from antiparallel pairs of pMOS transistors operated in accumulation.

The transistor level design of the 20 μS differential reference transconductor (Figure 5.14) started with the sizing of the nMOS and pMOS transistors to give equal transconductances of 20 μS while having a sufficiently low gate-overdrive voltage to ensure saturated operation with drain currents up to four times the quiescent current. For this design, the areas of the transistors were made large enough to give sufficient matching for the desired image rejection but not so large that their parasitic capacitance dominated the nodal capacitances required by the filter. The quiescent gate overdrive voltage was set to $V_{gt} \leq V_t/2 \approx 0.25$ V and, for the chosen process, this required the transistor sizing shown in Figure 5.17.

The circuit operates with $V_{dda} \approx 3\,V_t = 1.58\,V$ and transconductor drain currents of $J = 2.45$ μA. The calculated mismatch of the I- and Q-channel reference transconductors gave a transconductance standard deviation of $\sigma \approx 0.5$ per cent (the cross-branch transconductors will be better matched). The neutralisation network was designed with the appropriate feedback capacitance, $0.4(C_{gsp} + C_{gsn})$, and its source follower was biased sufficiently (0.35 μA) to keep it out of cut-off under large-signal, high-frequency conditions. The total current drawn by the differential reference transconductor including four half-sized common-mode feedback transconductors was 10.5 μA.

Next, the floating capacitors were designed. This was achieved by first sizing the MOS capacitors to give half the single-ended values defined in Table 5.1 and then including them with the designed transconductors in a transistor-level simulation

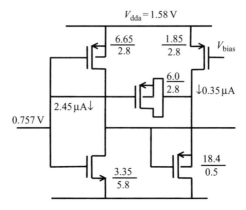

Figure 5.17   Transistor level design of 20 µS transconductor half circuit

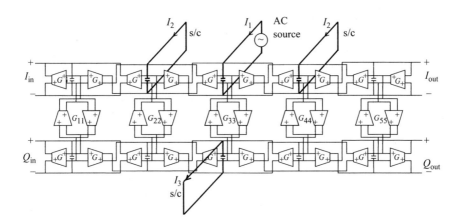

Figure 5.18   Filter arrangement for trimming nodal capacitances

model of the whole filter. Simulating this filter would produce a distorted amplitude response because each capacitor is loaded by the transconductor parasitic capacitances. So, each of the floating capacitors was trimmed to give the design value. Using the arrangement shown in Figure 5.18 with an AC voltage source connected across the selected capacitor (C_3 in this case) in the I-channel and short circuits across its neighbouring capacitors (C_2 and C_4) and across the corresponding capacitor (C_3) in the Q-channel, the selected capacitor together with its immediate parasitics are isolated from the rest of the filter network. The currents flowing in the short circuits allow extraction of the nodal capacitance ($C_3 = \text{imag}(I_1/2\pi f)$) and the surrounding transconductances ($G = \text{real}(I_2)$ and $G_{33} = \text{real}(I_3)$) for subsequent trimming to the design values. The same trimming procedure was adopted for each capacitor in turn. Typically, this trimming was only a few per cent of the capacitor value.

5.5 Performance

The reference transconductor was simulated to confirm its signal handling and the filter design was simulated to confirm its amplitude response, noise performance, signal handling, intermodulation distortion and rejection of power supply noise.

Under typical conditions, the single-ended transconductor (Figure 5.17) used transistors giving $g_{mp} = 19.94\,\mu S$ and $g_{mn} = 19.99\,\mu S$ and so $G = 39.93\,\mu S$. It consumes $2.8\,\mu A$ and has a quiescent input voltage of 0.756 V. When used in the differential transconductor and operated from an ideal 1.58 V supply rail, the simulated differential transconductance versus differential input voltage is as shown in Figure 5.19.

The amplitude responses and group delay are shown in Figures 5.20 and 5.21 and indicate that they are very close to ideal. Figure 5.22 shows the passband of the filter with no feedthrough equalisation and this indicates the severity of high-frequency peaking due to non-reciprocal feedthrough capacitances and the effectiveness of the feedthrough equalisation technique.

The 1 dB compression point occurred with a differential output swing of 1.3 V peak to peak (at 1 MHz), but the maximum differential voltage swing was restricted to 1 V peak to peak, corresponding to the maximum wanted Bluetooth signal of $-20\,dBm$ and required an input current swing of $40\,\mu A$ peak to peak. Figure 5.23 shows an output NPSD in the passband of aproximately $90\,nV/\sqrt{Hz}$. The signal-to-noise ratio, found by comparing the power of the maximum signal with that of the noise integrated over 100 MHz bandwidth, was 68.2 dB.

For the IM3 test, sinusoids at 4 MHz and 7 MHz with amplitudes 2 dB below the maximum signal were applied and the resulting simulated output spectrum is shown

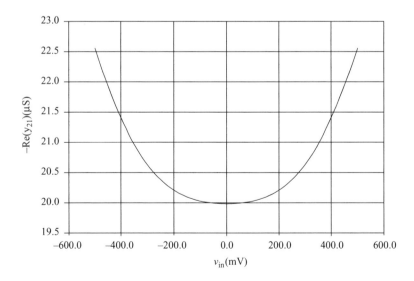

Figure 5.19   Simulated differential transconductance of 20 μS transconductor

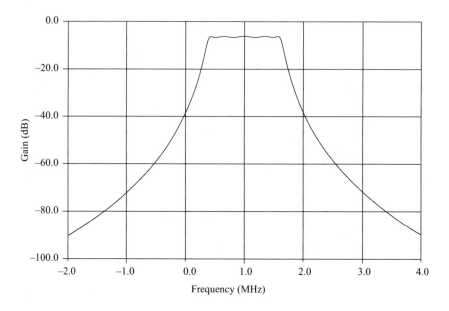

Figure 5.20   Simulated amplitude response

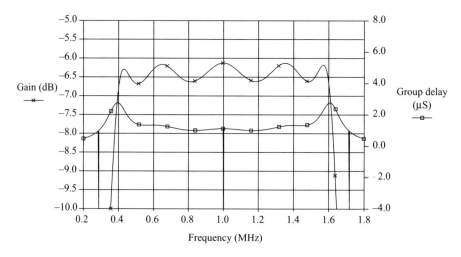

Figure 5.21   Simulated passband response and group delay

in Figure 5.24. The third-order product at 1 MHz was at a level of −86 dBV and this gives an IIP3 of 34.2 dBV.

For the supply noise intermodulation test, first V_{dd} and then V_{ss} were modulated by a 1.5 MHz sinusoid while the input was driven with a 0.5 MHz sinusoid, each with amplitudes of 100 mV peak. These signals were both chosen to be in the filter's passband because this gave greatest intermodulation. The simulated output

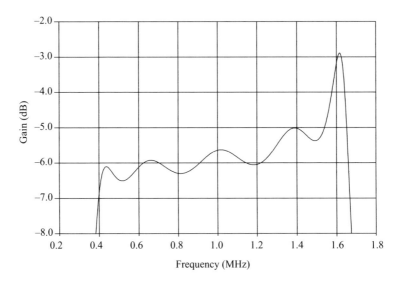

Figure 5.22   Simulated passband response without feedthrough equalisation

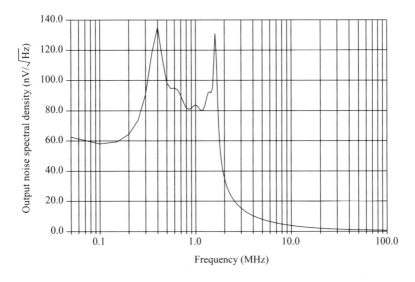

Figure 5.23   Simulated output noise spectral density

(nearly identical for either V_{dd} or V_{ss} excitation) is shown in Figure 5.25 and indicates intermodulation products at 1 MHz and 2 MHz at 47.7 dB and 48.7 dB below the direct signal at 0.5 MHz. This attenuation of about 48 dB is maintained at lower input signal levels and is even greater for out-of-band supply noise.

The total current drain for the whole Gm-C filter was 512 μA giving a power consumption of approximately 1 mW. The estimated chip area (including automatic

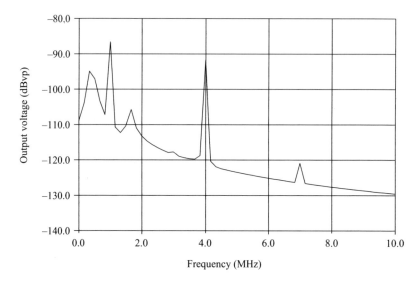

Figure 5.24   Simulated third-order intermodulation

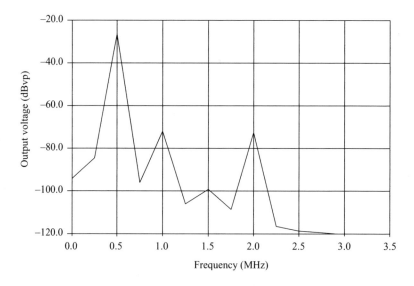

Figure 5.25   Simulated power supply-signal intermodulation

tuning) is approximately 0.18 mm^2. The results with nominal processing, at temperatures of $-20\,°C$, $27\,°C$ and $80\,°C$, are summarised in Table 5.2 and indicate good stability of performance.

With the transconductors deliberately skewed to emulate a 20 per cent mismatch between the transconductor's pMOS and nMOS transistors (the pMOS transistors

Table 5.2   Simulated performance of channel filter (nominal processing)

Process	2.5 V, 0.25 µm CMOS (C050FM)		
Filter shape	Chebyshev		
Filter order	5 + 5		
Filter ripple	0.5 dB		
Supply voltage (V_{dd})	2 V		
Temperature	−20 °C	27 °C	80 °C
Analogue supply(V_{dda})	1.568 V	1.580 V	1.607 V
Supply current (I_{dd})	410 µA	512 µA	641 µA
Centre frequency (F_0)	1.036 MHz	1.000 MHz	0.952 MHz
Bandwidth (F_{bw})	1.241 MHz	1.200 MHz	1.143 MHz
Gain (A_0)	−6.05 dB	−6.13 dB	−6.27 dB
Signal/noise (SNR)	68.6 dB	68.2 dB	67.9 dB
IIP3 (4 MHz, 7 MHz)	33.9 dBV	34.2 dBV	30.8 dBV
Supply intermodulation (0.5 MHz, 1.5 MHz)	−45.6 dB	−47.7 dB	−50.8 dB
Estimated chip area (filter)	$\approx 0.1\ \text{mm}^2$		
Estimated chip area (tuning)	$\approx 0.08\ \text{mm}^2$		

Table 5.3   Typical simulated performance of transconductor and channel filter with normal ($k_p/k_n = 1$) and skewed ($k_p/k_n = 1.2$) MOSTs

Transconductance ratio (k_p/k_n)	1.0	1.2
Transconductor		
Supply voltage (V_{dd})	2 V	2 V
Supply voltage (V_{dda})	1.580 V	1.586 V
Quiescent input voltage (V_{in})	0.756 V	0.770 V
Transconductance (G)	39.93 µS	40.57 µS
Channel Filter		
Centre frequency	1.000 MHz	1.014 MHz
Bandwidth	1.200 MHz	1.217 MHz
Passband gain	−6.13 dB	−6.27 dB
Signal/noise ratio	68.2 dB	68.2 dB
IIP3 (4 MHz, 7 MHz)	34.2 dBV	33.1 dBV
Supply intermodulation (0.5 MHz, 1.5 MHz)	−47.7 dB	−47.7 dB
Supply current (I_{dd})	512 µA	532 µA
Power dissipation	1.024 mW	1.064 mW

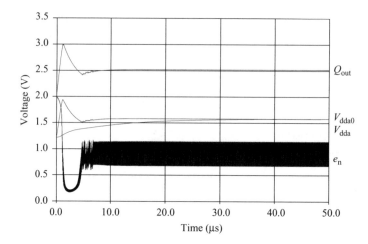

Figure 5.26   Simulated start-up behaviour of the tuning loop (see Figure 5.16)

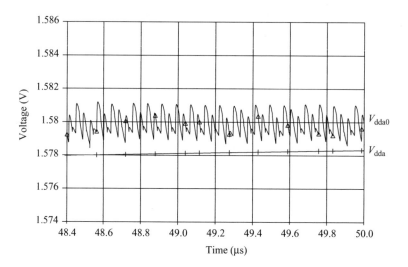

Figure 5.27   Simulated V_{dda} ripple of the tuning loop

were made 10 per cent wider and the nMOS transistors 10 per cent narrower), there were only minor changes to the typical performance as shown in Table 5.3. This justifies our claim that matching between the pMOS and nMOS transistor parameters is not critical.

Figure 5.26 shows the tuning loop (Figure 5.16) settling from start-up under typical conditions. It can be seen that the loop stabilises with the charge pump enable input, e_n, oscillating around 1 V and with its output, Q_{out}, rising above V_{dd} to about 2.5 V with V_{dda} settling to its correct value of 1.58 V. Figure 5.27 shows that the simulated ripple on V_{dda} is less than 0.1 mV. With the switched capacitor varied by ±40 per cent, the pole frequency of G/C stabilised to ±1 per cent. Over the temperature range from −20 °C to +80° C, the filter detuned by +3.6 per cent to −4.8 per cent. The output resistance of the V_{dda} regulator was about 80 Ω and the good rejection of V_{dd} and V_{ss} disturbances reported above is due in large part to the attenuation of these disturbances by the regulating transistor by about 20 dB. The power consumption of the tuning loop was about 100 μW which is about 10 per cent of the filter's power consumption.

5.6   Conclusions

A class AB polyphase Gm-C filter made in standard CMOS processing and dissipating only 1 mW has been presented. It uses a fully differential transconductor with novel feedthrough equalisation to make a balanced filter with near ideal response. It operates from a supply voltage regulator which automatically tunes the filter and provides effective rejection of both supply and substrate noise. The transconductors operate in class AB and this produces excellent efficiency and good linearity, making this filter technique a strong candidate for future wireless transceivers.

5.7   References

1 MINNIS, B. J., MOORE, P. A., PAYNE, A. W. and GREER, N. P. J.: 'A low-IF, polyphase receiver for DECT'. IEEE International Symposium on *Circuits and Systems*, 2000, pp. I60–I63

2 http://www.bluetooth.com

3 http://www.zigbee.com

4 SEDRA, A. S., SNELGROVE, W. M. and ALLEN, R.: 'Complex analog bandpass filters designed by linearly shifting real lowpass prototypes'. IEEE International Symposium on *Circuits and Systems*, 1985, pp. 1223–26

5 BEEHBAHANI, F., KISHIGAMI, Y., LEETE, J. and ABIDI, A.: 'CMOS mixer and polyphase filters for large image rejection', *IEEE Journal of Solid-State Circuits*, June, 2001, **36**, pp. 873–87

6 ALZAHER, H., ELWAN, H. and ISMAIL, M.: 'A CMOS highly linear channel select filter for 3G multistandard integrated wireless receivers', *IEEE Journal of Solid-State Circuits*, January, 2002, **37**, pp. 27–37

7 LIU, Q., SNELGROVE, W. M. and SEDRA, A. S.: 'Switched-capacitor implementation of complex filters'. IEEE International Symposium on *Circuits and Systems*, 1996, pp. 1121–24

8 WORAPISHET, A., SITDHIKORN, R. and HUGHES, J. B.: 'Low-power complex channel filtering using cascoded class AB switched-currents'. IEEE International Symposium on *Circuits and Systems*, 2002

9 HUGHES, J. B., WORAPISHET, A. and TOUMAZOU, C.: 'Switched-capacitors versus switched-currents: A theoretical comparison'. IEEE International Symposium on *Circuits and Systems*, 2000, pp. II409–12

10 MOULDING, K. W., QUARTLEY, J. R., RANKIN, P. J., THOMSON, R. S. and WILSON, G. A.: 'Gyrator video IC with automatic tuning', *IEEE Journal of Solid-State Circuits*, December, 1980, **15**, pp. 963–67

11 NAUTA, B. and SEEVINCK, E.: 'Linear CMOS transconductance element for VHF filters', *Electronics Letters*, 30th March, 1989, **25** (7), pp. 448–50

12 ANDREANI, P. and MATTISSON, S.: 'On the use of Nauta's transconductor in low-frequency CMOS Gm-C bandpass filters', *IEEE Journal of Solid-State Circuits*, February, 2002, **37** (2), pp. 114–24

13 TSIVIDIS, Y. P.: 'Operation and modeling of the MOS transistor' (McGraw-Hill, New York pp. 370–72)

14 SHI, B., SHAN, W. and ANDREANI, P.: 'A 57-dB image band rejection CMOS Gm-C polyphase filter with automatic frequency tuning for Bluetooth'. IEEE International Symposium on *Circuits and Systems*, 2002, pp. V-169–72

15 WU, J.-T. and CHANG, K.-L.: 'MOS charge pumps for low voltage operation', *IEEE Journal of Solid-State Circuits*, April, 1998, **33** (4), pp. 592–7

Chapter 6

Design and automatic tuning of integrated continuous-time filters

James Moritz and Yichuang Sun

6.1 Introduction

Examination of the block diagram of any wireless transceiver architecture reveals that filters are an essential building block. To satisfy all transceiver design requirements, many different types of filters operating over a wide range of frequency and bandwidth are required. Over a long period of time, several specialised technologies dedicated solely to filter implementation have evolved, for example, filters based on quartz crystal and ceramic resonators, LC filters using ferrites and other specialised magnetic materials, and at the upper end of the frequency range, transmission line elements fabricated as microstrip. These techniques yield high-performance filters; unfortunately, none of these components are available in designs using current IC technologies, and for a long time this created a barrier to the design of highly integrated RF systems, since many filtering functions had to be performed off-chip.

Developments in semiconductor processing and IC circuit design techniques during the past several years mean that it is now possible to implement continuous-time active filters with useful performance over an extremely wide frequency range. The possibility therefore exists of achieving the highly desirable goal of integrating all the filter functions required in many wireless transceiver applications. Single-chip transceivers with no external filtering components are now becoming commonplace for applications such as the Bluetooth and IEEE 802.11 standards [1, 2]. The current trend is to implement these designs using standard CMOS digital processes due to their low cost and ready availability. From the filter designer's point of view, a notable shortcoming of these current IC technologies is the loose tolerances that can be achieved on the values of on-chip components. Large variations in component values lead directly to large divergence of the achieved filter response from the intended design specifications. It is not normally possible to adjust component values after fabrication, therefore an almost universal requirement for integrated continuous time filters is the need for on-chip tuning, and the tuning system may well be the most difficult challenge in achieving satisfactory filter performance.

This chapter is divided into the following sections. Section 6.2 discusses requirements for filters in the context of fully integrated wireless transceivers, and the need for on-chip tuning systems. Section 6.3 describes methods of filter implementation. Different filter architectures including the second-order cascade, multiple loop feedback and ladder simulation are described. The most widely used techniques for filter implementation include the OTA-C and MOSFET-C techniques. The increasing operating frequency of integrated wireless transceivers and availability of on-chip inductors has led to the development of active-LC techniques. Section 6.4 discusses issues connected with filter tuning, including the difficulties introduced by circuit parasitics. Different methods of implementing filter tuning systems are outlined. Section 6.4.6 examines the special difficulties involved in tuning high-Q, high-frequency filters, and describes a proposed tuning technique for the leap-frog class of bandpass filter. The chapter is summarised in Section 6.5.

6.2 Filters in transceivers and the need for tuning

This section outlines the role of filters in typical wireless transceiver architectures, and some of the important parameters in their design. The need for on-chip tuning in virtually all practical applications of integrated filters is then discussed.

6.2.1 *Filter requirements for wireless transceivers*

A typical single-chip wireless transceiver design requires several different filters; possible architectures for a receiver and transmitter are shown in Figures 6.1 and 6.2 respectively [3]. In Figure 6.1, the signal input from the antenna is bandpass filtered at the signal frequency, typically in the VHF/UHF range. This defines the bandwidth of the receiver frontend, and reduces unwanted responses, such as the image response in the case of the superheterodyne architecture. Many designs rely on impedance

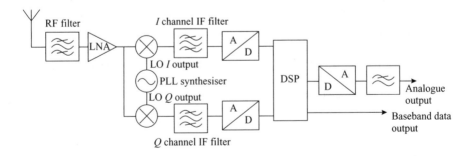

Figure 6.1   Typical receiver block diagram for wireless transceiver

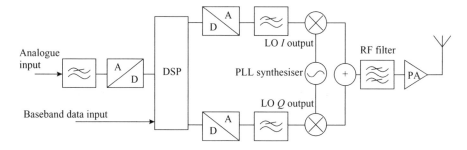

Figure 6.2   Typical transmitter block diagram for wireless transceiver

matching networks associated with the low-noise amplifier to provide sufficient selectivity, but in more demanding applications additional filters are required. The major challenge for integrated RF filter design is the high operating frequency; for example, GSM handsets operating around 900 MHz, Bluetooth at 2.45 GHz. Operation at these frequencies pushes the capabilities of current IC technologies, and circuit parasitics have a serious adverse effect on the response of the filter, making predictable performance difficult to achieve.

The signal is then down-converted to an intermediate frequency (IF), where further bandpass filtering occurs. Traditionally, the IF bandpass filter performs the channel filtering function before the signal is demodulated; in modern receiver designs this function is shared with subsequent digital signal processing. Each IF filter acts as an anti-aliasing filter for the succeeding converter or demodulator, and by providing rejection of unwanted signals in adjacent channels it reduces the dynamic range of signals applied to the ADCs. The bandwidth of the IF filters is defined by the type of modulation used; in multistandard receivers it may be necessary to provide a filter with selectable bandwidths. A low-IF architecture is shown in the example, with an image-rejection mixer. This architecture requires low-Q bandpass filters, with a centre frequency of similar order of magnitude to the bandwidth. Typical centre frequencies and bandwidths range from hundreds of kilohertz to several megahertz. As an alternative to the separate I and Q channel IF filters and ADCs shown, a single converter may be used in conjunction with a polyphase, complex bandpass filter which performs the image rejection function [4].

As with the other circuit functions in the receiver signal path, the filters will directly contribute noise and non-linearity. Thus as well as meeting specified requirements for amplitude, group delay, etc., the dynamic range of these filters is of critical importance.

The low-IF architecture is currently the most popular for wireless transceiver designs. Other receiver architectures may also be used. In the direct conversion ('zero IF') architecture, the incoming signal is converted directly to baseband, using a similar scheme to Figure 6.1, but with the IF bandpass filters replaced by low-pass filters. The cut-off frequency is then determined by the modulation bandwidth. The traditional superheterodyne design uses an IF frequency much larger than the signal bandwidth; this requires a high-Q, high-frequency, bandpass filter, and usually additional frequency conversion stages.

Many recent receiver designs digitise the received signal, and use a DSP 'back-end' to perform further signal processing and demodulation at low IF or baseband frequencies. Many wireless communications systems also require analogue outputs, for example audio signals. These require reconstruction filters at the DAC outputs.

The transmitter architecture of Figure 6.2 reverses the signal flow path of the receiver. The modulated signal is generated by DSP, and up-converted to the signal frequency. Again, anti-aliasing and reconstruction filters are required for the data converters. RF filtering at the output frequency is also required to suppress unwanted mixing products generated by the up-conversion process. The linearity and dynamic range requirements for filters in the transmit signal path are more relaxed than in the receiver because the transmitter operates over a relatively narrow range of signal levels.

Therefore, we see that a typical transceiver contains several filters operating over a very wide frequency range, from input frequencies in the gigahertz range to audio reconstruction filters. Several different active filter techniques are often used within the same transceiver to suit these differing requirements. These are discussed in more detail in Section 6.3.

6.2.2   The need for on-chip tuning

On-chip tuning is essential for most filters because of the extremely wide tolerances of components fabricated in the IC processes normally used for the fabrication of the mixed-signal 'system on a chip' (SoC). In terms of volume, the vast majority of ICs being manufactured currently are digital in nature, and as a result, CMOS processes optimised for digital logic are the most highly developed and economical for high-volume, low-cost production. Because most wireless transceiver applications are very cost-sensitive, and the analogue RF parts of these transceivers represent only a small proportion of the total circuitry on the chip, an overwhelming economic case exists for using CMOS digital processes for the implementation of the complete transceiver. From the viewpoint of the analogue designer, CMOS processes, whilst yielding active devices capable of excellent high-speed performance, are mediocre from the point of view of component value accuracy. Each type of component is fabricated during several process steps, each of which contributes variability to the final component value.

Typically, initial component tolerances of the order of several per cent to tens of per cent can be expected. It is interesting to note that, in spite of the great strides made in semiconductor technology during the past few decades, component tolerances have improved little [5]. It is possible to fabricate accurate on-chip components, but the need for extensive additional processing makes this impractical for applications where high yield and low cost are of paramount importance. The pole and zero frequencies of the filter are dependent on the values of at least two different types of component, for example, resistance and capacitance or transconductance and capacitance, each of which is subject to uncorrelated tolerance variations. Therefore, tolerances on the initial cut-off or centre frequencies of the filter of 50 per cent or more can be expected.

In principle, the desired filter response could be obtained by trimming components on the die after fabrication. However, this is usually impractical for economic reasons. In standard IC processing, a large number of chips are processed in parallel, as a single operation, simultaneously on a single wafer. This is a fundamental factor that allows production of complex ICs at very low cost. On the other hand, a post-fabrication trimming process would have to be performed on each die individually, adding significantly to test times and so to production cost. Also, even after trimming, the components on the IC are subject to further substantial variations in value due to the effects of ageing and environmental variables, in particular temperature.

Therefore, a viable fully integrated filter design for a wireless transceiver requires some form of on-chip automatic tuning system. The filter itself is built from electrically tuneable voltage- or current-controlled elements, and tuning control circuits adjust the tuning signal values until the required response to a reference signal is obtained. This firstly requires filter circuit structures that are electrically tuneable over a sufficient range to compensate for component tolerances, whilst at the same time capable of meeting all system design requirements for filter response, dynamic range, etc. Suitable circuit techniques that have been exploited include g_m-C, MOSFET-C and active-LC; these are discussed in Section 6.3. Secondly, a tuning system must be devised which is capable of detecting whether the filter response is correct within acceptable limits, and if not, apply suitable corrections to the tuning signals to restore correct operation. Because changing chip operating conditions lead to significant changes in filter performance, the tuning process must usually proceed continuously, or periodically at frequent intervals during normal operation of the chip, without upsetting the operation of the system as a whole. A wide range of on-chip tuning methods have been devised; some of these are discussed in Section 6.4

6.3 Filter architectures and design methods

As we have seen in Section 6.2, the most common types of filter response required are lowpass and bandpass, but occasionally a demand arises for other types; highpass, band-reject and allpass, together with equaliser functions. The components available for filter design depend on the IC technology used; for current designs, pure CMOS is by far the most popular technology, and further discussion will focus on CMOS circuit techniques. Bipolar technologies have been extensively used also, and offer some performance benefits in that bipolar devices have greater transconductance and lower noise for a given bias current. From the designer's viewpoint, BiCMOS would appear to offer the best choice of implementations. All these technologies offer resistors and capacitors with good performance, however inductors are only useable for applications in the gigahertz range. To design lower frequency filters with complex poles and zeros in their response without inductance requires the use of active circuits with gain. The other necessary requirement for integrated filter designs is tunability, which requires at least one type of component to be made variable. This

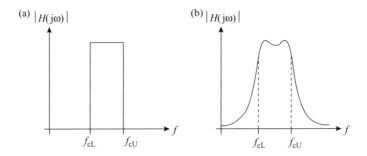

Figure 6.3   (a) Ideal bandpass filter response; (b) practically realisable response

can be achieved by using a component with a value dependent on bias voltage or current. An alternative to electrically tuneable circuit elements is the use of switched arrays, in which component variation in discrete steps is achieved. This is a technique well suited to digital control; it is also useful for extending the tuning range.

At the system level, the required response of a filter may be specified in a number of ways. The designer is usually interested in the amplitude response with changing frequency, and also the response in the time domain. This latter can be expressed either in terms of phase response, or group delay versus frequency. An idealised bandpass filter response is shown in Figure 6.3a; here the amplitude response is unity inside the passband, but zero beyond the upper and lower cut-off frequencies f_{cL} and f_{cU}. The same filter would have a constant group delay (or a linearly increasing phase lag) with changing frequency. In practice, as in Figure 6.3b it is only possible to realise filter responses that approximate to the ideals; different approximations attempt to optimise one aspect of filter response, or to achieve the best compromise between a number of conflicting requirements. Thus many different approximating functions have been devised; the Chebyshev and elliptic functions aim to approach the ideal 'brick wall' amplitude response, the Butterworth response achieves maximum gain flatness within the passband, and the Bessel and linear phase responses approximate a constant group delay. Procedures are available to select the optimum filter response for a particular application from a catalogue [6]. Numerical methods can also be used to optimise a practically achievable filter function to meet a specific set of requirements [7].

The response of a continuous-time filter is usually specified as a polynomial function of the complex frequency operator s:

$$H(s) = \frac{N(s)}{D(s)} = \frac{a_n s^n + a_{n-1} s^{n-1} + \cdots + a_1 s + a_0}{s^m + b_{m-1} s^{m-1} + \cdots + b_1 s + b_0}. \qquad (6.1)$$

In order to be practically realised, restrictions are that $m \geq n$, a_i, b_i are real, and b_i are positive [8]. This function can be factorised to give:

$$H(s) = a_n \frac{\prod_{i=1}^{n}(s - z_i)}{\prod_{j=1}^{m}(s - p_j)} \qquad (6.2)$$

where p_j are the poles of $H(s)$, that is, the roots of $D(s)$, and z_i are the zeros of $H(s)$, the roots of $N(s)$. In general, z_i and p_j are either real, or exist in complex conjugate pairs. The filter response is therefore defined as a set of poles and zeros. Once the designer has selected a suitable response of this form, several different filter circuit architectures are available for implementation. The most important architectures are the second-order cascade, multiple loop feedback structures, and simulations of passive LCR ladders. These are discussed in the following sections.

6.3.1 Second-order cascade architecture

Any transfer function of the form (6.2), and of order n, can be factorised into a series of N second-order responses of the form:

$$t_i(s) = \frac{a_{i2}s^2 + a_{i1}s + a_{i0}}{s^2 + b_{i1}s + b_{i0}} \tag{6.3}$$

where $n = 2N$. Where the order n is odd, an additional first-order function $t(s)$ is required. The complete transfer function can then be expressed as

$$H(s) = t(s). \prod_{i=1}^{N} t_i(s) \tag{6.4}$$

where $t(s) = 1$ if the transfer function is even order. This can be physically realised as a cascade of circuit blocks, each implementing a second-order response, with a single first-order section if required. The stucture is shown in Figure 6.4. Each second-order block is often called a 'biquad', due to the biquadratic form of equation (6.3). Each biquad implements two complex conjugate poles, and none, one or two zeros, depending on requirements. The design of the biquads must be such that the input of each biquad can be connected to the preceding output without loading effects.

A vast range of biquad circuits have been devised [8, 9]. Single amplifier biquads (SABs) consist of a resistor–capacitor network with an amplifier embedded in it. Although SABs have the advantage of lower power consumption due to having only a single active device, they are little used for integrated filters. This is because SABs have higher sensitivity to component tolerances and integrated passive components require a larger chip area than active devices. Also, tuning the transfer function of most SABs is more difficult than is the case for two integrator loop biquads, since there are several different types of components used in the circuit, each subject to large, uncorrelated mismatches. Other biquads are based on a feedback loop containing a pair of integrators. One possible two-integrator loop configuration is shown in Figure 6.5; this is a representation of the Tow–Thomas active RC biquad.

Figure 6.4 *Second-order cascade structure*

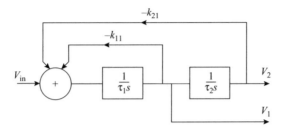

Figure 6.5   Block diagram of Tow–Thomas biquad

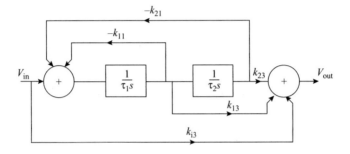

Figure 6.6   Biquad with output summation

The transfer functions at the outputs of this block diagram are:

$$\frac{V_1}{V_{in}} = \frac{(1/\tau_1)s}{s^2 + (k_{11}/\tau_1)s + k_{21}/(\tau_1\tau_2)}, \quad \frac{V_2}{V_{in}} = \frac{1/(\tau_1\tau_2)}{s^2 + (k_{11}/\tau_1)s + k_{21}/(\tau_1\tau_2)}.$$

(6.5)

The transfer function at V_2 is a lowpass function with two complex conjugate poles; a bandpass function is available at V_1, with a zero along with the same pole pair. This simple biquad is therefore restricted to lowpass and bandpass filters based on all-pole filter responses. More complex implementations of the two-integrator loop allow the designer to include arbitrary zeros in the filter response. One method to obtain zeros in the biquad response is by taking a weighted sum of the input and outputs of Figure 6.5, as shown in Figure 6.6. An alternative approach is to inject the input signal into all the internal nodes of the biquad as in Figure 6.7.

Both these circuits can implement the arbitrary second-order filter function of equation (6.3) by suitable choice of the coefficients k_{ij}; for Figure 6.6:

$$\frac{V_{out}}{V_{in}} = \frac{k_{i3}s^2 + ((k_{i3}k_{11} + k_{13})/\tau_1)s + (k_{i3}k_{21} + k_{23})/\tau_1\tau_2}{s^2 + (k_{11}/\tau_1)s + k_{21}/\tau_1\tau_2}$$

(6.6)

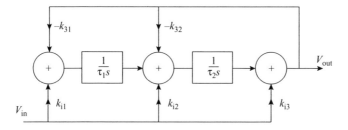

Figure 6.7   Biquad with input distribution

and for Figure 6.7:

$$\frac{V_{out}}{V_{in}} = \frac{k_{i3}s^2 + (k_{i2}/\tau_2)s + (k_{i1}/\tau_1\tau_2)}{s^2 + (k_{32}/\tau_2)s + (k_{31}/\tau_1\tau_2)}.$$ (6.7)

The circuit of Figure 6.7 is especially desirable, since the coefficients controlling zeros and poles are completely independent of each other.

The complete second-order cascade filter is designed in modular form, using a selection of biquads suitable to realise the required poles and zeros in the filter response. Any response can be realised in many different ways; in principle, the set of biquads could be cascaded in any order without changing the overall response. However, considerable design effort is required to optimise the design with respect to signal-to-noise ratio and dynamic range. Pole–zero pairing is important in optimising the dynamic range of each biquad [8–10]; if the pole and zero pairs in any one biquad are at widely different frequencies, the gain of the biquad will vary widely over the operating frequency range. This restricts the range of input signals which can be applied, since a large signal will cause overloading at frequencies where the gain is high, while small signals applied at the minimum-gain frequencies will have poor signal-to-noise ratios. The poles and zeros are therefore usually distributed between the biquads so as to minimise the frequency separation. The sequence in which the biquads are cascaded has a major effect on the dynamic range of the filter. If a biquad with a large-amplitude peak in its response is included as the first stage of the cascade, subsequent stages may be overloaded by signals close to this peak frequency. As a general rule, this results in biquads being cascaded in ascending order of Q factor [9].

The second-order cascade offers the benefits of ease of design and flexibility. This configuration is also easy to tune, since each second-order block operates independently of others in the cascade, and so can be tuned separately without mutual interaction. However, the sensitivity of the second-order cascade is higher than for the other types of filter implementation described below. Also, at high frequencies, it is difficult to achieve sufficient isolation between biquads to avoid loading effects and interdependence of tuning.

6.3.2   *Multiple-loop feedback architecture*

The principle of the two-integrator loop can be extended to a cascade of a larger number of integrators with multiple feedback paths. A high-order filter function can then be realised directly without dividing it into second-order sections. The advantage this approach brings is that sensitivity is reduced [8, 11].

A general multiple loop feedback (MLF) filter configuration is shown in Figure 6.8. It consists of a cascade of integrators, with feedback paths which may connect between any of the integrator outputs and inputs. For a high-order filter containing several integrators, a very large number of permutations of feedback path configurations exist. Normally, only a small proportion of the possible feedback paths are implemented, since this leads to a simplified design without reducing the functionality of the circuit.

Special cases of the general MLF structure are the follow-the-leader feedback (FLF) structure of Figure 6.9, where all integrator outputs are summed at the input of the first integrator, and the IFLF (inverse follow-the-leader feedback) structure of Figure 6.10, where feedback to all integrator inputs is taken from the final integrator

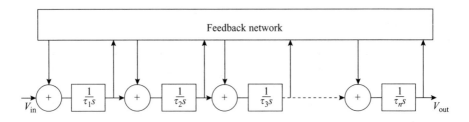

Figure 6.8   General MLF filter structure

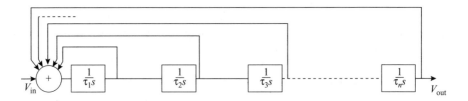

Figure 6.9   FLF structure

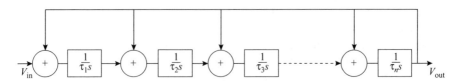

Figure 6.10   IFLF structure

output in the cascade. A further important special case is the 'leap-frog' structure, used to simulate LC ladder filter structures; this is discussed in the following section.

As in the case of the simple two-integrator loops of the previous section, the MLF structures above realise frequency responses containing only poles. Also as in the case of the second-order circuits, input distribution and output summation can be used to introduce zeros in the response.

The advantage of MLF filter configurations in general is that they offer lower sensitivity than the second-order cascade. Drawbacks are that because all components in the filter interact to generate the filter transfer function, tuning any single component will affect all the poles and zeros of the response. This makes on-chip tuning of these filters relatively difficult; in practice, this difficulty is usually avoided using the master–slave technique described in Section 6.4.3. High-frequency parasitics give rise to problems with MLF filters having feedback paths which extend around a large number of integrators, such as the FLF and IFLF above. An ideal integrator will have a phase shift between output and input of exactly 90°. In real integrators, parasitic circuit effects will give rise to an actual phase shift which leads or lags slightly the nominal 90°. This 'excess phase' gives rise to severe Q-enhancement effects as will be discussed in Section 6.4.1, and cascading several integrators within any one feedback loop will lead to an accumulation of excess phase. Parasitic feed-forward via the feedback paths also leads to distortion of the filter transfer function.

6.3.3 LC ladder simulation

Many design techniques have been developed for passive LCR ladder filters. These filters have low sensitivity, so are attractive for integrated filter implementation, but usually not practical, except at very high frequencies, because of the very restricted availability of integrated inductors. However, circuit techniques are available which simulate the operation of LC filters using active devices, and as a result share the same design techniques and low sensitivity. The component substitution approach to LC ladder simulation uses active circuits which simulate the behaviour of inductance to substitute for actual inductors. A different approach is to simulate the voltages and currents in the passive LC circuit using active integrators. Both these approaches lead to broadly equivalent results.

The component substitution technique begins with a passive LCR design, such as the lowpass filter of Figure 6.11. The inductors are then replaced by a two-terminal circuit which has the same relationship between voltage and current as the inductors it replaces. One such circuit using a gyrator composed of operational transconductance amplifiers, in conjunction with a capacitor, is shown in Figure 6.12.

Simple analysis shows that the impedance between the terminals of this circuit is:

$$\frac{V_{in}}{I_{in}} = \frac{sC}{g_1 g_2}. \qquad (6.8)$$

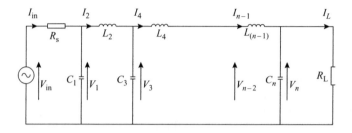

Figure 6.11   Lowpass LC ladder

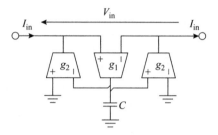

Figure 6.12   Simulated inductance

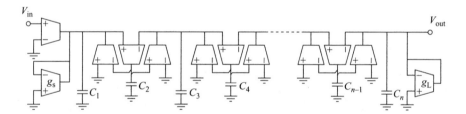

Figure 6.13   Ladder filter using gyrators

Thus the circuit behaves as an inductance $L = C/g_1 g_2$. Replacing all the inductors in Figure 6.11 with the gyrator circuit with suitably dimensioned values of transconductance and capacitance will clearly result in an active filter with the same transfer function as the prototype. The resulting circuit is shown in Figure 6.13.

Another feature of the circuit of Figure 6.13, in addition to the elimination of inductors, is that the impedances in the series branches of the ladder have been transformed into shunt impedances with one terminal connected to ground. Modifying the circuit topology in this way is often useful, since it simplifies implementation of the circuit; in this case, grounded capacitors are less subject to parasitic effects. A range of similar circuit transformations can be used to modify the filter to produce the most practical and economic realisation, for example to minimise the numbers of active elements required. The inductance simulating gyrator circuit has used operational transconductance amplifiers as an example, and is a popular circuit technique

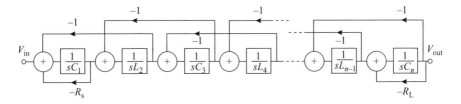

Figure 6.14   Leap-frog structure

(see Section 6.3.4). However, many other circuit techniques can also be used to realise gyrators; for example op-amp-RC techniques [8].

A different approach to realising an active circuit with the same transfer function as a passive ladder circuit is to simulate the signals at different points within the ladder circuit. The voltages across the shunt arms, and currents in the series arms of the LC ladder prototype of Figure 6.11 are given by

$$I_s = \frac{1}{R_s}(V_{in} - V_1), \quad V_1 = \frac{1}{sC_1}(I_s - I_2), \quad I_2 = \frac{1}{sL_2}(V_1 - V_3),$$

$$V_3 = \frac{1}{sC_3}(I_2 - I_4), \dots,$$

$$I_{(n-1)} = \frac{1}{sL_{(n-1)}}(V_{(n-2)} - V_n), \quad V_n = \frac{1}{sC_n}(I_{(n-1)} - I_L). \tag{6.9}$$

Each of these equations is the integrated sum of two voltages or currents; a circuit which generates the same overall transfer function can therefore be made up of integrators and summers, as in Figure 6.14.

This circuit is a special case of the multiple loop feedback circuits of the previous section; due to the feedback paths between alternate integrators, it is called the 'leap-frog' (LF) configuration. In Figure 6.11 the signals within the circuit are a mixture of voltages and currents. However, in practice it is more convenient to scale the signals such that all signals are either voltages or currents [12, 13].

Because these types of filter are derived from passive LCR prototypes, the design techniques developed for the passive filters can be utilised to generate active filters via simulation techniques. Further scaling of gain and impedance is also required to give practical component values, and to optimise dynamic range, as in the case of the second-order cascade. Although simple all-pole lowpass filters have been used in the illustrations above, it is possible to simulate any type of response that can be realised using LCR ladder filters [8, 13]. By modifications to the basic LF structure, it is also possible to realise responses that are not possible with the passive ladder structure, such as equalisers [14]. Although the inductor substitution and LF ladder simulation have been presented as different techniques, it can be shown that the two approaches give equivalent results, and yield the same circuit under certain conditions. Compared to the FLF and IFLF multiple loop feedback structures, the feedback paths in LF structures only enclose two integrators, making them less affected by excess phase in the integrators and so more suitable for high-frequency applications.

6.3.4   OTA-C filters

OTA-C filters utilise two types of component – operational transconductance amplifiers (OTA), and capacitors. Filter behaviour is determined by the open-loop transconductance of the OTAs, hence these are also known as g_m-C filters. The OTA operates as a voltage-controlled current source, with differential inputs; ideally, it has infinite input and output impedances and the output current is linearly proportional to the differential input voltage. The basic building block for filter design is the integrator shown in Figure 6.15a. The transfer function of this integrator is

$$\frac{V_{out}}{V_{in}} = \frac{g_1/g_2}{1 + sC/g_2},$$

(6.10)

i.e. the circuit is a lossy integrator. If g_2 is eliminated, a lossless integrator results:

$$\frac{V_{out}}{V_{in}} = \frac{1}{sC/g_1}.$$

(6.11)

Since ideally the OTA outputs have infinite impedance, and so are unaffected by the output voltage, the summing function in OTA-C filter structures is performed simply by paralleling the outputs of two or more OTAs.

In practice, fully differential circuits such as in Figure 6.15b are almost universally used. The same transfer function is achieved, but the symmetrical structure has important advantages for noise rejection. Provided symmetry is maintained in the differential circuit, noise will be coupled equally into each differential pair of nodes. Thus, the noise will give rise to common-mode signals, but the symmetrical nature of the differential OTAs ensures a high degree of common-mode signal rejection. This is extremely important in mixed-signal IC design, where a high level of noise may be generated by digital circuits sharing the same substrate. A further benefit of differential structures is the cancellation of even-order distortion products, improving linearity.

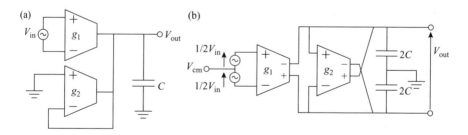

Figure 6.15   (a) g_m-C integrator, and (b) balanced g_m-C integrator

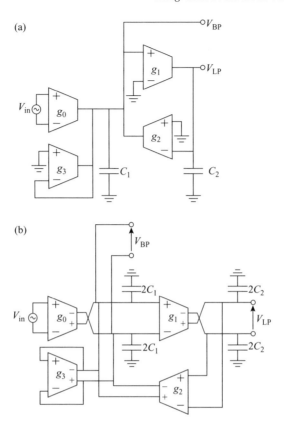

Figure 6.16   (*a*) *Single-ended Tow–Thomas biquad;* (*b*) *balanced Tow–Thomas biquad*

An example of a g_m-C biquad is shown in Figure 6.16, based on the Tow–Thomas active-RC biquad. Figure 6.16a shows a single-ended circuit, whilst Figure 6.16b is a balanced circuit with the same transfer function. Lowpass and bandpass filter functions are generated simultaneously:

$$H_{LP}(s) = \frac{V_{LP}}{V_{in}} = \frac{-g_0 g_1}{s^2 C_1 C_2 + s g_3 C_1 + g_1 g_2} \tag{6.12}$$

$$H_{BP}(s) = \frac{V_{BP}}{V_{in}} = \frac{s g_0 C_1}{s^2 C_1 C_2 + s g_3 C_1 + g_1 g_2} \tag{6.13}$$

$$\omega_0 = \sqrt{\frac{g_1 g_2}{C_1 C_2}}, \quad Q = \frac{1}{g_3}\sqrt{\frac{g_1 g_2 C_2}{C_1}}, \quad K_{LP} = \frac{g_0}{g_2}, \quad K_{BP} = \frac{g_0}{g_3}. \tag{6.14}$$

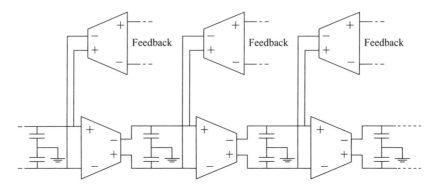

Figure 6.17   MLF g_m-C filter structure

A feature of this and OTA-C filter circuits in general is that all the nodes in the filter circuit have capacitors connected to ground. This has the important advantage for high-frequency operation that no parasitic poles will be introduced by the effects of stray capacitance between these nodes and the substrate, which is at AC ground potential; instead, the stray capacitance will simply increase the total capacitance at that node. Capacitors fabricated on an IC have significant capacitance between the plates and the substrate, especially the bottom plate which is closest to the substrate. Where all capacitors have the bottom plate terminal grounded, no AC current flows into this stray capacitance. The top plate-substrate capacitance, which is much smaller, is absorbed into the total capacitance. Furthermore, the OTA itself normally has no high-impedance internal nodes. Therefore, the transfer function of the OTA-C structure has minimal parasitic poles, minimising problems caused by excess phase shift. The OTA-C technique is therefore well suited to high-frequency applications [15].

An example of a MLF OTA-C filter structure is shown in Figure 6.17. Normally, a higher-order filter structure is formed from a cascade of several integrators. Since the integrators have high input impedances, they may be directly cascaded. The feedback paths are implemented with additional OTAs.

An elementary CMOS OTA using a differential pair is shown in Figure 6.18a. The transconductance is tuned by varying the bias current I_D. However, this elementary OTA has severe performance drawbacks; in particular, poor linearity and low output impedance. At the same time, attempting to tune the transconductance over a wide range degrades performance, increasing noise and reducing bandwidth. Much effort has been expended on producing OTAs with a usefully wide tuning range while at the same time minimising non-ideal behaviour.

In OTA-C integrators, in contrast to the closed-loop op-amp based designs, the OTA operates in an open-loop mode. Any non-linearity in the voltage-to-current transfer function of the OTA will give rise to distortion products in the output signal from the integrator. Therefore, a great deal of design effort has been expended on making the voltage-to-current conversion of the OTA as linear as possible. The main

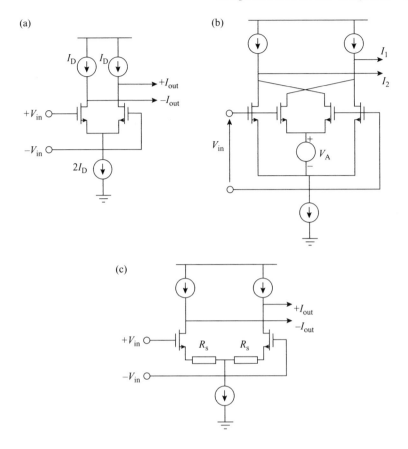

Figure 6.18   Simple CMOS OTAs

techniques used to improve linearity are non-linear term cancellation, and source degeneration [16–18].

The concept of the first technique is to sum the outputs of a number of non-linear transconductors in such a way that the non-linear terms cancel, leaving a residual linear transconductance. An example of this technique is the cross-coupled quad shown in Figure 6.18b. Assuming the drain current of the MOSFETs operating in the saturation region obeys a square-law relationship, i.e. $I_D = K(V_{GS} - V_T)^2$, it can be shown that $I_{out} = I_2 - I_1 = 2K V_A V_{in}$. Thus the composite transconductance of the quad of MOSFETs is perfectly linear, and is also tuneable by varying V_A. In practice, linearity will not be perfect, due to mismatches between devices making up the quad, and also because of imperfect compliance with the square law I_D/V_{GS} relationship. Noise will also increase. Similar linearization schemes have been developed based on the exponential I_C/V_{BE} relationship in bipolar transistor OTAs [19].

In the source degeneration technique, local negative feedback is applied to the OTA input devices. This is accomplished by the source resistors R_s in Figure 6.18c.

This reduces the dependence of transconductance on input signal level, and so reduces distortion. However, it also reduces the dependence of transconductance on the bias current, so drastically reducing the range over which the transconductance can be tuned. To obtain a useful tuning range, the degeneration resistors are replaced by MOSFETs operating in the triode region. A number of circuit configurations are possible [16]. The MOSFETs are effectively voltage-controlled resistors, and the gain of the OTA is tuned by varying their gate bias voltage. The MOSFET channel resistance is itself a non-linear function of the applied signal voltage, so optimising the linearity of the source-degenerated OTA requires careful design and modelling. Adding source degeneration inevitably reduces the gain of the OTA, therefore to achieve the same transconductance as the non-degenerated OTA, larger transistors, higher bias current and greater chip area is required.

The other major defect of the simple OTA circuit is the finite output impedance. This is due to the dependence of the MOSFET drain current on the drain-source voltage. The output impedance can be greatly increased by isolating the input MOSFETs from changes in voltage at the output terminals. Three techniques for achieving this are shown in Figure 6.19.

A simple cascode output is shown in Figure 6.19a; this has the drawback of reducing the input and output voltage swing that is possible while maintaining the MOSFETs in their active operating region. The folded cascode circuit of Figure 6.19b allows much larger signal voltage excursions. It also allows the output voltage to be more negative than the input voltage, so that no additional level shifting is required when a number of such OTAs are cascaded. The current mirror circuit of Figure 6.19c achieves a similar result to the folded cascode; additionally, more complex types of current mirror can be used to increase the output impedance further.

Further output transistors can be added to the current mirror of Figure 6.19c to provide multiple, identical outputs. These may be used to reduce the number of OTAs required in a filter; for example, in the MLF structure of Figure 6.17, the OTAs in the feedback paths could be eliminated by taking feedback from duplicate outputs of the integrator OTAs. A similar economy can be achieved by using OTAs with multiple pairs of inputs [17, 20]; this can be achieved by an OTA circuit with multiple input stages, the output being produced by summing the input stage currents.

6.3.5   MOSFET-C filters

MOSFET-C filters are extremely widely used for lower-frequency applications where low distortion is important. The MOSFET-C technique was derived from active-RC techniques; in discrete component designs, the useful frequency range of the active-RC filter is limited by parasitic capacitances to frequencies mostly below 1 MHz. However, in a fully integrated design, circuit parasitics can be reduced sufficiently to allow satisfactory performance at frequencies of 100 MHz or greater [15, 21]. As with other types of integrated filter, it is necessary to make components electrically tuneable to correct for component value variations.

To achieve this, the resistors of the active-RC design are replaced with MOSFETs operating in the triode region, where the channel acts as a variable resistance which

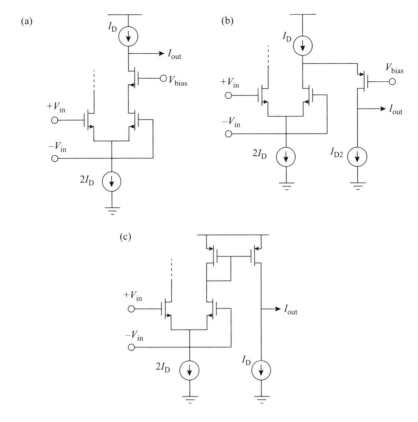

Figure 6.19   Increasing OTA output impedance: (a) cascode, (b) folded cascode and (c) current mirror

is dependent on the gate and substrate bias voltages. The cut-off frequency and Q of the filter is tuned by varying the gate voltages. A MOSFET used in this way is usually called a MOS resistor. The basic active-RC integrator is shown in Figure 6.20a. Figure 6.20b shows a basic MOSFET-C integrator. Within the operating frequency range of the filter, the closed loop gain of the integrator is much smaller than the open-loop gain of the amplifier, so the usual benefits of high loop gain in reducing distortion are obtained. This makes it unnecessary for the amplifier to have a very linear transfer function, as is the case for gm-C filters. This also allows a large excursion of the input and output voltages to be utilised, maximising the upper limit of dynamic range, an important factor with the trend in falling supply voltages.

Practical MOSFET-C filters invariably use a balanced integrator structure, such as that shown in Figure 6.20c, due to Banu and Tsividis [22]. As well as the usual noise-reduction advantages of balanced circuits, this achieves the advantage of a high degree of cancellation of unwanted distortion products generated by the inherently non-linear MOS resistors. With small input signal levels, each op-amp input behaves as a virtual

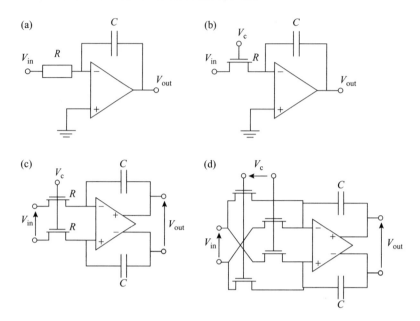

Figure 6.20   Active-RC and MOSFET-C integrators

earth, and each MOS resistor as a linear resistor, as with the active *RC* integrator. With large input signals, the MOS resistor channel conductance is modulated by the applied signal voltage, and substantial, mostly second-order distortion products are present at the op-amp inputs. However, even-order distortion products are largely cancelled by the differential nature of the circuit. Thus while individual components of the MOSFET-C integrator are highly non-linear, the overall circuit can have high linearity, an example of externally linear, internally non-linear circuit behaviour.

The range over which the MOS resistors can be tuned in value depends on the magnitude of input signals and the range of gate control voltage available. High input signal levels will lead to the MOS resistors operating outside the triode region in some parts of the input waveform, causing increased distortion. A modification of the MOSFET-C integrator structure has been described by Czarnul [23]. As shown in Figure 6.20(d), an additional pair of cross-connected MOS resistors is added to the input of the integrator, which partially cancel the input current. The input structure is similar to that of a double-balanced mixer. The integrator gain is dependent on the ratios of the MOSFET resistances, and can be varied over a wide range by varying the differential bias voltage applied to the MOSFET gates. This circuit in principle has advantages in improving linearity, and also reducing the effects of parasitic MOSFET resistor capacitance at high frequency, although device mismatches reduce the improvement which can actually be achieved.

To a first-order approximation, the MOSFET-C integrator is unaffected by the parasitic capacitances associated with the capacitors and MOS resistors. The op-amp

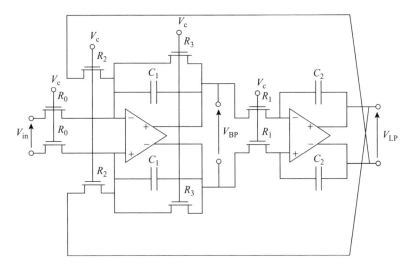

Figure 6.21 MOSFET-C biquad

inputs are a virtual ground and so little AC current flows in the stray capacitance between the op-amp inputs and ground. Capacitances between the op-amp outputs and ground are driven by the low-impedance outputs; provided the loop gain is still sufficiently high, the transfer function of the integrator is unaffected. However, the input voltage and output impedance of the op-amp are finite, so stray capacitance will in fact produce parasitic poles at high frequencies. Also the open-loop gain of the amplifier will decrease at high frequencies.

A MOSFET-C biquad is shown in Figure 6.21. This is an implementation of the two integrator loop of Figure 6.5; the summing function is performed at the virtual earth nodes at the integrator inputs, by providing extra MOS resistors (R_2, R_3) in the feedback paths. Lowpass and bandpass functions are available from the circuit:

$$\frac{V_{\text{LP}}}{V_{\text{in}}} = \frac{R_2}{R_0} \frac{1/(C_1 C_2 R_1 R_2)}{s^2 + s/C_1 R_3 + 1/(C_1 C_2 R_1 R_2)},$$

$$\frac{V_{\text{BP}}}{V_{\text{in}}} = \frac{R_3}{R_0} \frac{s/C_1 R_3}{s^2 + s/C_1 R_3 + 1/(C_1 C_2 R_1 R_2)}. \tag{6.15}$$

The same basic structure can be extended to higher-order MLF filters in a similar way to the g_{m}-C structure of Figure 6.16.

The amplifier component of the MOSFET-C integrator may be a low output impedance operational amplifier, or it may be an operational transconductance amplifier with large g_{m}, such that a high loop gain is achieved with the output loaded. Other types of amplifier may also be employed [24]. Achieving sufficient gain requires a multistage amplifier design, with frequency compensation to ensure adequate phase margin for closed-loop operation. A further requirement is for differential outputs

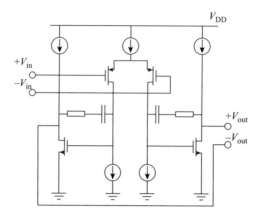

Figure 6.22   Balanced op-amp

to implement the balanced integrator structure. A typical balanced op-amp circuit is shown in Figure 6.22.

6.3.6   Active-LC filters

Classical passive LC filters have many advantages in signal processing; high operating frequencies, wide dynamic range and zero power consumption. However, the inductors required are difficult to integrate. On-chip inductors take the form of flat spirals using the upper metalisation layers of the IC, and become prohibitively large for values greater than a few nanohenries. But current IC technology has now extended the upper frequency limits of integrated transceivers well into the gigahertz range, where filter design with inductance values in this range is practical. The new limitation is the low Q that can be achieved; optimised inductor design gives Q values in the range 5–20 [15, 25]. This severely limits the range of filter responses which can be achieved. In spite of the low Q factors, integrated inductors are an essential component in many integrated transceiver functions, such as low-noise amplifiers, voltage controlled oscillators, and power amplifiers, so considerable research effort is being expended on improving integrated inductor performance.

The effective Q of integrated inductive components can be increased by cancelling the resistive losses of the inductors with active negative resistance generating circuits. In a passive network, Q may be defined as the ratio of the reactive to the resistive parts of the impedance of a component. Figure 6.23a shows a simple model of a lossy inductor with a Q of 6 at 2 GHz. The losses in the inductor are modelled as a series resistor. Clearly, if a circuit simulating a negative resistance of $-8.4\ \Omega$ were connected in series with Figure 6.23a, then the losses of the inductor would be perfectly compensated. A parallel negative resistance R_p could also be used to cancel out the inductor losses at 2 GHz:

$$R_p = -\frac{R_s^2 + \omega^2 L_s^2}{R_s},$$

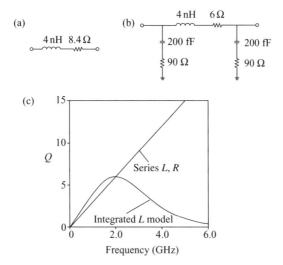

Figure 6.23   *Simple inductor model, and more realistic models of integrated spiral inductor*

giving a value of $-372 \ \Omega$. Practically, it is easier to generate this higher value of negative resistance, since active components of lower transconductance are required. However, a parallel negative resistor only gives exact compensation of series loss resistance at a single frequency, due to the dependence on ω of the effective parallel resistance. The behaviour of the integrated planar spiral inductor is substantially altered by large parasitics, in particular the distributed capacitance between the spiral metalisation and the substrate, and eddy currents induced in the substrate.

A more accurate equivalent circuit for a typical spiral inductor is shown in Figure 6.23b (from Reference 26). This component also has a Q of 6 at 2 GHz. The Q of this component, and that of the simple series model, are plotted against frequency in Figure 6.23c. No single compensating negative resistance can be connected to the network of Figure 6.23b that will exactly compensate the loss over a wide frequency range. Therefore, it is practically only possible to achieve Q compensation over a fairly narrow frequency range, and active-LC circuits are primarily of interest as narrowband bandpass filters. Several filters giving useful performance as RF bandpass filters have been described [27–29]

Circuits that generate negative resistance rely on positive feedback; for example the transconductance amplifier configuration of Figure 6.24a, in which

$$Z = \frac{V}{I} = \frac{V}{-g_{\mathrm{m}}V} = -\frac{1}{g_{\mathrm{m}}}. \tag{6.16}$$

OTA designs used for other types of active filter could be used for the transconductance in Figure 6.24a, however, due to the very high operating frequency of active-LC filters, the more complex amplifiers used for lower frequency applications are impractical

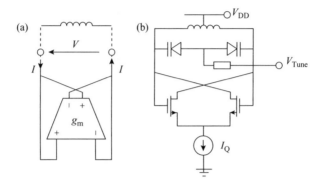

Figure 6.24   (a) Negative resistance generator and (b) active LC filter section

due to their bandwidth limitations. Very simple tranconductance amplifier circuits, similar to those used for voltage controlled oscillators, are generally used, as in Figure 6.24b.

When the inductor is combined with a resonating capacitor, a second-order bandpass biquad results. Varying the bias current I_Q in the active devices controls the transconductance g_m, and therefore the negative resistance and Q of the filter. The resonant frequency of the filter is largely determined by the inductance and capacitance values. No means currently exist of tuning an integrated inductor, so the capacitance must be made variable to achieve frequency tuning. This can be accomplished using varactor diodes; unfortunately, these have rather limited tuning range, and also intro-duce additional losses and non-linearity into the circuit. Another method of tuning the resonating capacitor is to use switched capacitor arrays as will be described in Section 6.3.7. The main difficulty in implementing the switched array is to make the losses and parasitic capacitances in the array switches sufficiently small.

High-order active-LC filters are derived from passive LC prototypes. Integrated inductors with useably high Q are limited to a narrow range of values, so the best implementation is a coupled resonator structure as shown in Figure 6.25a, which can be designed using identical inductors. The shape of the filter response is defined by the Q of each LCR resonator, and the coupling coefficient between them. In the case shown, coupling between L_1 and L_2 is provided by the mutual inductance M_{12}. The value of M_{12} is defined by the geometrical layout of the inductors, which is little affected by process variations. Therefore, inductors with the required value and coupling coefficient can be fabricated with relatively close tolerances, and the need for tuning the coupling coefficient is avoided.

A typical fourth-order bandpass active-LC filter is shown in Figure 6.25b. The loss in the two resonators of the prototype (R_{Q1}, R_{Q2}) is replaced by the net resistive loss in L_1, L_2, after partial compensation by the negative resistance generating circuits. I_{Q1} and I_{Q2} therefore control the Q of the resonators, and therefore the filter response shape. The filter centre frequency is tuned by the varactor diode bias voltages V_{tune1} and V_{tune2}.

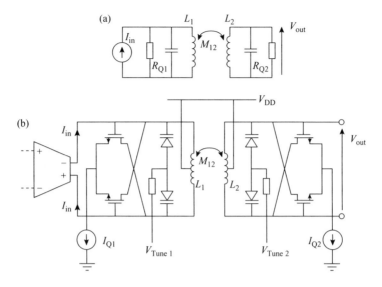

Figure 6.25 (a) Coupled resonator prototype, and (b) active LC fourth-order bandpass filter

6.3.7 Switched arrays for filter tuning

Electrically tuned components exploit the dependence of a component parameter on a bias current or voltage, for example the dependence of g_m on bias current for a transistor, the resistance of a MOS resistor on gate voltage, or the capacitance of a varactor diode on the reverse bias voltage. Electrical tuning is well suited to tuning circuits using analogue techniques and where high tuning resolution is required. Drawbacks are that the changing bias levels usually affect other aspects of the device performance: linearity, noise and parasitic components. This restricts the tuning range that can be achieved; typically, a 2:1 tuning ratio is obtained.

Switched arrays consist of an array of similar components, usually with a binary sequence of values, that are switched in or out of the circuit to produce an overall value which can be adjusted in discrete steps. Switched arrays have the advantage of making normally fixed components, such as passive resistors and capacitors, tuneable. It is possible to make the component value tuneable over a wider range than is practical with electrically tuned components.

An example is the switched capacitor array shown in Figure 6.26. A value of $0-31C$ in steps of C can be obtained by applying the appropriate digital control word. It is common to make each larger value from a combination of unit value capacitors, each with an identical switch, rather than capacitors and switches with scaled areas. This improves the matching between sections because all capacitors are subject to similar fringing and stray capacitances. Also, the series resistance of the switches scales exactly with the capacitance, maintaining the same parasitic time constant with different values of capacitance selected.

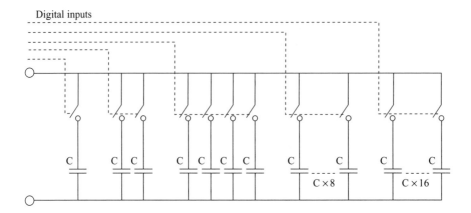

Figure 6.26   Switched capacitor array

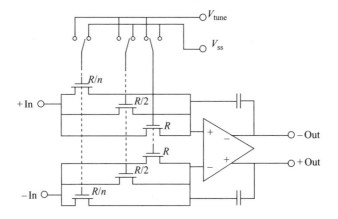

Figure 6.27   MOSFET-C integrator with switched resistors

The switched array technique is especially suited to the MOS resistor, as shown in Figure 6.27; several MOSFETs are connected as a parallel array in which the devices being used have the control voltage applied to their gates. The resistors not required are deselected by biasing them into cut-off. Thus no additional switching devices are required in the signal path of the filter, reducing circuit parasitics. This technique is widely applied to MOSFET-C filters; it can also be applied to OTA-C filters using source degeneration. Transconductance can also be tuned using the switched array technique; an OTA is constructed using scaled current mirrors as the output stage. Individual current mirrors may be biased on or off by the digital control word, giving a binary scaled output current.

The switched array can be considered a form of digital-to-analogue converter, and so is inherently suited to tuning schemes using digital techniques. Switched capacitor arrays are used where extended tuning range is required, for example in multistandard transceiver designs, where a wide range of bandwidths may be required to suit different standards using the same circuit. However it is difficult to achieve high tuning resolution due to the large number of elements required, and the difficulty in accurately matching a wide spread of values in the array and between different arrays. Also, performance is degraded by switch parasitics, such as on resistance and stray capacitance. For these reasons, a combination of electrical tuning and switched arrays may be used for wide-range, high-resolution tuning [20].

6.4 On-chip tuning of continuous-time filters

6.4.1 *Frequency tuning and Q tuning*

As discussed in Section 6.2, on-chip tuning of integrated filters is generally required due to the wide tolerances of integrated components. The passband and cut-off frequencies of a particular filter response are defined by the time constants in the circuit, which are in turn determined by the component values, i.e. *RC*, *LC* products or ratios C/g_m. Since the absolute tolerance of each type of component is quite large, and there is generally no correlation between the errors in value of different types of component, the overall tolerance of the pole and zero frequencies is larger still – of the order of 50 per cent in typical IC processes.

In contrast to the very loose absolute tolerances of integrated components, ratio matching between components of the same type on a single die is quite accurate. Normally, ratios are maintained to better than a few per cent, and may be within 0.1 per cent when special layout techniques are used. The shape of the filter response, defined by the Q of the poles and zeros, and mid-band gain, K, of a filter are defined primarily by ratios between similar components. Since these are well defined, it would seem that Q and gain would automatically also be defined accurately. This can be true for filters operating at moderate frequencies and with moderate Q, but high-Q designs are severely affected by parasitic effects, in particular, phase shifts caused by high-frequency parasitic poles and zeros in the amplifiers, and finite DC amplifier gain.

These factors are illustrated in the tuneable second-order OTA-C section shown in Figure 6.28. The OTA-C biquad of Section 6.3.4 is used here as an example, although the same considerations apply equally to other filter techniques. The circuit generates simultaneous lowpass and bandpass outputs; here we focus on the bandpass transfer function, but the lowpass case is very similar. The transconductances in the filter are made tuneable by varying their bias currents. The capacitors are fixed.

In the case where all components are ideal, the transfer function of Figure 6.28 is

$$H_{\text{BP}}(s) = \frac{V_{\text{BP}}}{V_{\text{in}}} = \frac{g_0}{g_3} \frac{(g_3/C_2)s}{s^2 + (g_3/C_2)s + (g_1g_2/C_1C_2)}. \tag{6.17}$$

By equating coefficients with the standard form of the second-order bandpass transfer function

$$H_{\text{BP}}(s) = K_{\text{BP}} \frac{s(\omega_0/Q)}{s^2 + s(\omega_0/Q) + \omega_0^2}, \tag{6.18}$$

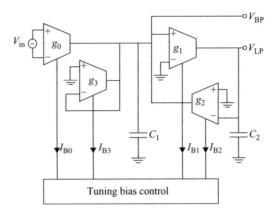

Figure 6.28   Tuneable OTA-C biquad

we have

$$\omega_0 = \sqrt{\frac{g_1 g_2}{C_1 C_2}}, \quad Q = \frac{1}{g_3}\sqrt{\frac{g_1 g_2 C_2}{C_1}}, \quad K_{BP} = \frac{g_0}{g_3}. \tag{6.19}$$

Suppose processing variations cause the transconductances of the OTA to vary. Because of the inherent matching between the components making up the OTAs, all transconductances will change by the same factor k_g. Similarly, process variations will cause all the capacitor values to change by a factor k_c, which in general will be different from, and unrelated to, k_g. Including the factors k_g and k_c in (6.17) gives a new transfer function:

$$H'(s)_{BP} = \frac{g_0}{g_3} \frac{(k_g g_3/k_c C_2)s}{s^2 + (k_g g_3/k_c C_2)s + (k_g^2 g_1 g_2/k_c^2 C_1 C_2)}. \tag{6.20}$$

ω_0 has been changed by a factor k_g/k_c. In order to restore the design value of ω_0, the four OTAs are simultaneously tuned until $k_g = k_c$, in which case (6.20) reduces to (6.17). In practice, this is achieved by tuning the transconductances until ω_0 is equal to the design value. It is also possible to tune Q independently of ω_0 by varying g_3 alone. However, because Q is determined by the relatively accurately defined ratios of C_1 to C_2, and g_1, g_2 to g_3, it would appear that Q would remain virtually constant during the tuning process.

In a real design, however, circuit parasitics can have a major effect on the Q of the circuit of Figure 6.28. An ideal OTA has transconductance that is independent of frequency, but real OTAs have finite bandwidth, caused by parasitic high-frequency poles and zeros associated with circuit nodes inside the OTA. Although these normally occur far above the passband frequency of the filter, and have little effect on the magnitude of integrator gain within the passband of the filter, they increase the phase shift of the integrators to slightly more than the nominal 90°. This 'excess phase', as illustrated in Figure 6.29b, can substantially alter circuit Q.

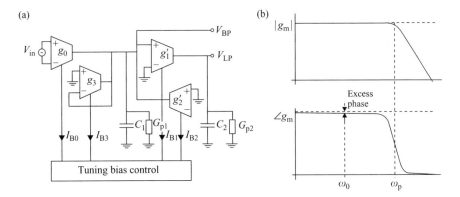

Figure 6.29 (a) Non-ideal biquad and (b) excess phase

In the frequency range of interest, where $\omega \ll \omega_p$, the frequency-dependent OTA gain $g'(s)$ can be modelled as an ideal transconductance g with an added phase shift proportional to frequency:

$$g'(s) = g\frac{1}{1 + s/\omega_p} \cong g\left(1 - \frac{s}{\omega_p}\right) \cong g\exp\left(-\frac{s}{\omega_p}\right). \tag{6.21}$$

In the circuit of Figure 6.29a, the most significant influence of excess phase occurs in the two integrators made up of g_1, C_2 and g_2, C_1. Substituting this frequency dependent transconductance for the ideal transconductors in (6.17) and making appropriate approximations for $\omega_0 \ll \omega_p$ gives a new value of Q for the circuit when excess phase is included:

$$Q' \approx \frac{Q}{1 - 2Q\omega_0/\omega_p}. \tag{6.22}$$

Q' is significantly affected by quite small values of excess phase. For example, if the design value of Q is 10 and $\omega_p = 100\omega_0$, giving rise to excess phase of about $0.6°$, Q' from (6.22) is 12.5, an increase of 25 per cent. A design Q of 50 will result in Q' approaching infinity, and instability.

Another parasitic effect that modifies Q is the finite output conductance of OTAs. The output of an ideal OTA behaves as a current source, but real OTAs have significant output resistance. This can be modelled by conductances G_{p1} and G_{p2} shunting each internal circuit node to ground, as shown in Figure 6.29a. The transfer function of this modified circuit is given by:

$$H(s)'' = \left[\left(\frac{g_0 G_{p2}}{C_1 C_2}\right)\left(1 + \frac{C_2}{G_{p2}}s\right)\right] \Big/ \left[s^2 + \left(\frac{G_{p2}}{C_2} + \frac{G_{p1}}{C_1} + \frac{g_3}{C_1}\right)s\right.$$

$$\left. + \left(\frac{G_{p1}G_{p2}}{C_1 C_2} + \frac{G_{p2}g_3}{C_1 C_2} + \frac{g_1 g_2}{C_1 C_2}\right)\right]. \tag{6.23}$$

The Q of the modified circuit is approximated by:

$$Q'' = \frac{Q}{1 + \frac{Q}{\omega_0}\left(G_{p1}/C_1 + G_{p2}/C_2\right)}. \tag{6.24}$$

Thus, increasing output conductance of the OTAs reduces Q. This sets an upper limit to the Q which can be achieved for a given set of transconductors and capacitors; as the design Q in (6.19) tends to infinity, the maximum achievable Q'' is

$$Q''_{max} = \frac{1}{\frac{1}{\omega_0}\left(G_{p1}/C_1 + G_{p2}/C_2\right)}. \tag{6.25}$$

In summary, the large tolerances of integrated components give rise to large frequency errors in the filter response, so integrated filters almost always require on-chip tuning. Frequency tuning will often be sufficient for filters operating at modest values of Q and frequency, typically lowpass and bandpass filters where the bandwidth is of the same order as the centre frequency. However in high-Q, high-frequency filters, circuit parasitics, principally the excess phase and finite DC gain of the active circuits, profoundly affect the Q of the filter response, so that Q must be tuned also.

6.4.2   On-line and off-line tuning

The outline of a typical tuning system is shown in Figure 6.30. A well-defined reference signal is applied to the filter input. One or more parameters of the filter output signal are measured by the frequency tuning control circuit and compared to a reference. The resultant error signal is used by the control circuit to calculate a correction signal, which is then applied to the frequency tuning input of the filter. Thus the system forms a closed feedback loop in which the filter is forced to converge on the desired frequency response. In a similar way, if implemented, the Q control circuit generates a tuning signal that corrects the Q of the filter.

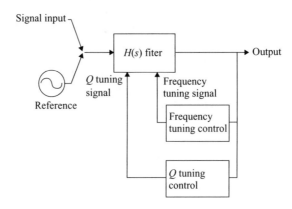

Figure 6.30   Outline of on-chip tuning scheme

Desirable features of any on-chip tuning system are minimal chip area and low power consumption. This requires simple hardware, and minimal computation requirements for the tuning algorithm. Conversely, the functional requirements placed on the filter design may be very complex, requiring several different performance goals for cut-off frequencies, gain, group delay, etc., which must be simultaneously met. It is not usually possible for an on-chip tuning system to evaluate all the relevant parameters of filter performance since this requires many measurements to be performed on the filter output signal over a range of frequencies. In high-order filters, there are many tuneable components, so achieving the desired filter response requires the control of a large number of variables simultaneously. For these reasons it is very difficult to directly tune a high-order filter using reasonably simple tuning circuits.

In order for the filter to function correctly, the tuning system must operate when the chip is first powered on. Also, component values will continuously drift while the circuit is powered, due to changes in environmental and operating conditions, so it is necessary to periodically repeat the tuning process during normal operation. This creates a problem in that the reference will be present within the filter passband at the same time as the desired signal, with the inevitable possibility of mutual interference occurring between the tuning system and the rest of the transceiver signal processing. The scheme of Figure 6.30 is therefore normally operated as an 'off-line' tuning system; periodically the normal signal input to the filter is removed, and the reference signal applied. The filter is then tuned, and the updated values of frequency and Q tuning signals stored until the next tuning cycle occurs.

These off-line tuning cycles can readily be accommodated in many transceiver architectures; for example, many transceivers alternate between transmit and receive; receiver filter tuning can take place during the transmit periods without affecting receiver operation. However, the additional signal routing and the requirement to store the tuning signals while the filter is on-line lead to added complication. Therefore, 'on-line' tuning is widely used where the tuning process proceeds continuously, and simultaneously with normal circuit operation. One way of achieving this is to devise a reference signal that has minimal effect on subsequent signal processing, but which at the same time can be used to measure the necessary filter parameters. An example of this is described in [30], where the reference signal is made nearly orthogonal to the received signal using spread-spectrum techniques.

6.4.3 Master–slave tuning

A very widely used and important on-line tuning scheme is the master–slave tuning scheme outlined in Figure 6.31. This makes use of the inherent good matching between components and circuit subsystems that are achieved within a single IC. Two well-matched filter sections are used; the reference signal is applied to the master section, whilst the actual input signal is applied to the slave section. The tuning system develops tuning signals in a closed feedback loop which correct errors in the response of the master section, as in off-line tuning. The same tuning signals are simultaneously applied to the slave section. If the master and slave sections are identical and perfectly matched, the response of the slave filter will be the same as the master. Thus

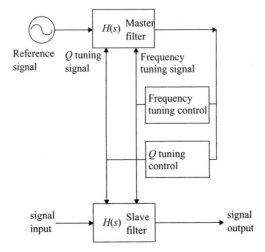

Figure 6.31   Master–slave tuning scheme

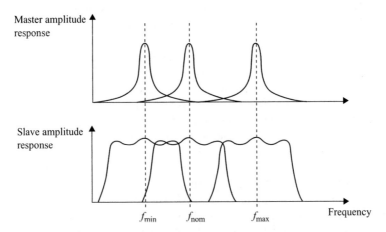

Figure 6.32   Tuning range of master and slave filter sections

it is unnecessary to apply the reference signal to the master filter, which can operate continuously.

In practice, master and slave are usually different. The master section is usually a low-order filter, often a biquad, since this has a simple response for which it is relatively easy to design tuning algorithms and is economical of chip area and power consumption. This is illustrated in Figure 6.32; the master filter is a single biquad with a single resonance peak in its response. The slave section can be of whatever order is required to meet the filter specifications, with the same tuning signals applied to each section. The diagram shows the effect on the frequency responses of master and slave as they are simultaneously tuned. Clearly, it is much easier to design a tuning algorithm for the single biquad master when compared to the high-order slave response, with its multiple maxima and minima. Thus in addition to allowing on-line tuning, the master–slave tuning scheme provides a solution to the problem of tuning complex filters. A large proportion of high-order integrated filter designs therefore utilise master–slave tuning in some form.

The essential assumption made in the master–slave scheme is that the ratios of components in the master and slave sections can be accurately achieved, and will track each other precisely as the master section is tuned. If this is the case, the cut-off frequencies and Q of all the filter sections in the slave will exactly track those of the master, and if the master section Q is maintained at the correct value, the slave filter response shape will remain correct as the filter frequency is tuned. There are practical limitations on how nearly this can be achieved, and a substantial amount of design and layout effort must be expended to ensure that the master accurately models the tuning behaviour of the slave. Since parasitic effects can substantially alter the performance of the filter, these must also be accurately modelled in the master section. These requirements can usually be best met by designing the slave filter with circuits that are as near identical to the master as possible, and by making the tuning reference signal frequency close to that of the signal frequency. This allows the best matching between sections, and also ensures that frequency-dependent parasitic effects are similar in master and slave. Synthesis techniques are required that result in filter circuits using the minimum possible spread of component values.

6.4.4 Frequency tuning methods

For frequency tuning, the most commonly used input reference signal is derived from a stable clock oscillator. This is convenient, since most systems already include an accurate off-chip clock signal, usually derived from a quartz crystal, from which all on-chip clock signals are derived through various forms of frequency synthesis. At the output of the filter, phase comparison is the most widely used method of determining the state of tuning of the filter. An outline frequency tuning scheme is shown in Figure 6.33.

In second-order filter sections, the phase difference between filter input and output reach well-defined values at the resonance frequency; 90° in the case of a lowpass

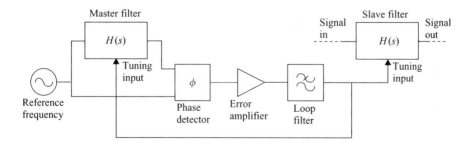

Figure 6.33 *Outline frequency tuning scheme*

section, $0°$ for a bandpass section. Accurate reference phase shifts independent of component value tolerances can be generated using digital counters operating on multiples of the reference frequency. An example of a phase detector is an analogue multiplier or mixer. With two inputs of the same frequency applied to the inputs, a DC output voltage is generated which is a function of the phase difference of the input signals. This transfer function is non-linear, and also depends on the signal amplitudes; but if the phase difference is arranged to be $90°$, the output will be zero. Therefore, if the output of the phase detector is applied to the frequency tuning input of the filter via an error amplifier and loop filter, feedback around the loop will cause the cut-off frequency of the filter to reach a value giving the desired value of phase shift.

This system is, however, subject to several sources of error. With large initial errors in filter resonant frequency, the reference signal is likely to be outside the filter passband at the start of the tuning procedure. A filter section with high Q and therefore rapid cut-off in amplitude will attenuate the reference signal to a low level that may prevent the correct operation of the phase detector. This limits the scheme to low-Q filter sections. A low-Q filter has a phase response that changes only slowly as the frequency is tuned, therefore for a given tuning error, only a relatively small error signal is produced at the phase detector output. Therefore the signal-to-noise ratio inside the tuning loop is poor; small errors due to DC offsets in the phase detector, or parasitic phase shifts, give rise to large tuning errors. A further source of errors is distortion in the reference or filter output waveforms; applying a non-sinusoidal signal to the phase detector may result in incorrect output.

A modification of this method that avoids these problems is to replace the filter section being tuned by a voltage controlled oscillator (VCO). This can be achieved by applying feedback with the appropriate gain and phase to the filter section. Oscillation then occurs at the resonant frequency of the filter. The output signal frequency is compared with the reference frequency using a phase detector, and the error signal is used to tune the VCO as before. This system is therefore a phase-locked loop, as shown in Figure 6.34. The advantage of using the output frequency of a VCO rather than the phase shift through the filter section as the basis for tuning is that the system is independent of fixed errors in the phase comparison; provided the phase difference between reference and VCO is maintained constant, the VCO frequency, and therefore the resonant frequency of the filter, is identical to the reference.

The main source of tuning error in this system is the mismatch in behaviour between the filter section, operating at finite Q and with relatively low signal levels, and the VCO which effectively operates at infinite Q and inherently requires a non-linear amplitude limiting mechanism to achieve a stable signal amplitude. To ensure that the frequency-determining elements operate within their linear range, the VCO is usually implemented by adding a limiting amplifier to provide feedback around a bandpass biquad filter section.

Many successful frequency tuning systems using frequency- or phase-locked loops as described have been implemented in practical designs, for example References 31 and 32. These methods are well suited to master–slave designs, where the tuning loop can operate continuously. This yields an extremely simple control system, and is often capable of frequency tuning accuracy within 1 per cent . These

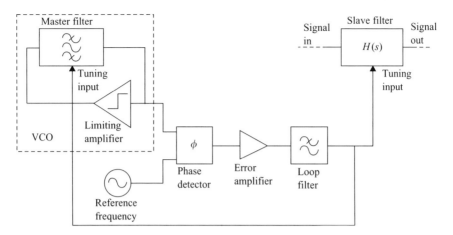

Figure 6.34   PLL frequency tuning system

techniques become increasingly difficult to apply at the highest frequencies, due to the increasingly severe errors caused by excess phase, both in the filter or VCO and in the phase detector itself.

Frequency tuning techniques can utilise the time-domain response of the filter. The response of a high Q bandpass filter to a step or impulse function is a damped sinusoid at the filter output. The period of the sinusoid is approximately related to the resonant frequency of the filter. The filter output waveform is squared using a limiting amplifier, and the period measured using digital counter techniques. A tuning signal is derived by comparing the measured period with the desired value. In order to achieve good accuracy, high resolution in the period measurement is necessary. This requires that the transient response has a long duration. The duration increases with Q and filter order, and due to this and the iterative nature of the measurement technique it is most appropriate for off-line tuning of high-order, high-Q bandpass filters [33, 34]. This tuning control method is digital in nature, so is easily combined with switched array tuning schemes.

A related technique is to measure the time constant of an integrator using a DC charging current. An example of this technique using an OTA-C integrator is shown in Figure 6.35.

An accurate clock signal is used to open the switch for a period t. During t, the integrator output voltage is a linear ramp that reaches a maximum value of

$$V_{out}(\max) = V_{ref}t\frac{g_m}{C}. \tag{6.26}$$

This maximum voltage is stored by the peak detector, and compared with the reference voltage by the error amplifier. The resulting low-pass filtered error signal is applied to the OTA transconductance control input and causes the capacitor charging current and therefore $V_{out}(\max)$ to vary. Over a large number of clock cycles, this feedback

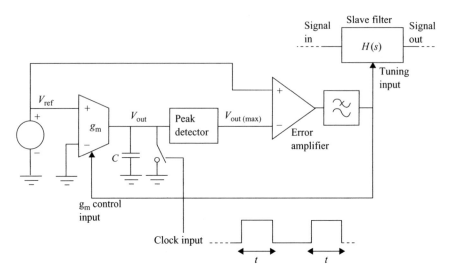

Figure 6.35   Frequency tuning scheme based on ramp generator

loop causes $V_{out}(\text{max})$ to become equal to V_{ref}:

$$V_{ref}t\frac{g_m}{C} = V_{ref}, \quad \text{so} \quad \frac{g_m}{C} = t. \tag{6.27}$$

Since t is accurately defined by the clock signal, and the resonant frequency of the filter is accurately proportional to g_m/C due to well-defined ratios between all transconductances and capacitances on the chip, the resonant frequency is set to the correct value.

In order to avoid problems caused by unwanted phase shifts, frequency tuning methods have been devised based on amplitude measurements. A second-order response with Q greater than $1/\sqrt{2}$ contains a peak in its amplitude response plotted against frequency. For high Q values, the amplitude peak closely approximates the resonant frequency. Tuning the resonant frequency of the filter with a fixed input reference frequency will also produce a peak in output response when the two frequencies coincide. The tuning system only needs to detect when the maximum output signal is achieved; the amplitude detector need therefore have neither high accuracy nor linearity, provided it has a monotonic response.

A tuning scheme using this principle is shown in Figure 6.36. The reference signal is applied to the biquad input, and the envelope detector produces a DC level, V_{env}, proportional to the amplitude of the filter output. In the first phase of the tuning cycle, the filter tuning voltage V_{tune} is swept through its range by the ramp generator. At the point where the resonant frequency of the filter coincides with the reference frequency, the filter output amplitude and thus V_{env} reaches a maximum, and this value is stored by the peak detector as V_{pk}. In the second tuning phase, V_{tune} is swept again, and V_{env} is compared with V_{pk} by the comparator. At the point where the resonant frequency

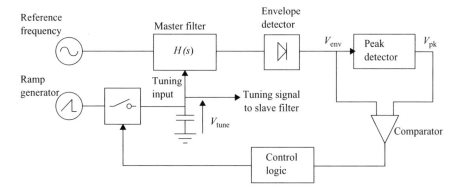

Figure 6.36   Frequency tuning based on amplitude peak detection

and reference frequency coincide, V_{env} is equal to V_{pk} and the control logic opens the switch. Thus the value of V_{tune} giving the correct filter resonant frequency is stored on the hold capacitor, until the next tuning cycle begins.

In practice, the circuit of Figure 6.36 will suffer tuning errors due to parasitic charge injection and loss from the tuning voltage holding capacitor, and offsets in the comparator and peak detector. However, a more sophisticated implementation of this technique has been described [35] in which these errors are largely eliminated.

6.4.5   Q tuning techniques

Frequency tuning ensures that the centre frequency or cut-off frequency of the filter is tuned to the correct value; however this does not necessarily ensure that the shape of the frequency response is correct; this also depends on the Q of the filter sections. As noted above, parasitic effects in particular may lead to severe distortion of the filter response. The frequency tuning schemes described above are independent of Q. However, in order to tune the filter Q, it is first necessary that the frequency tuning process is completed, because Q is defined in terms of the way that filter response changes close to the resonant frequency. Any error in the filter resonant frequency will therefore also result in errors in Q. A further difficulty is that although the designer will attempt as far as possible to make Q unaffected by frequency tuning and vice versa, they are never entirely independent. Inevitably, tuning Q will introduce a new error in filter frequency, and correcting this error will alter Q again.

Therefore, several iterations of frequency and Q tuning may be required to correctly tune the filter, or the two processes must proceed simultaneously. The designer must take the interdependence of both tuning processes into account in order to ensure that convergence takes place [36]. This is especially difficult with high-Q filter sections, where as seen in Section 6.4.1, Q is sensitive to small changes in tuning, and instability can easily occur.

The most widely used Q tuning technique [35, 37] utilises the fact that in many cases the gain of a biquad at the resonant frequency is proportional to the Q.

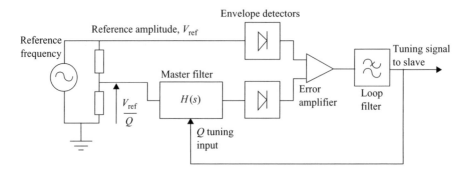

Figure 6.37   Q tuning scheme

For example, in the case of the OTA-C biquad of Figure 6.16, from equations (6.12), (6.13) and (6.14), we can derive expressions for the gain in terms of Q at ω_0:

$$\left|\frac{V_{LP}}{V_{in}}\right| = Q, \quad \left|\frac{V_{BP}}{V_{in}}\right| = Q\frac{g_0}{\sqrt{g_1 g_2 C_2/C_1}}. \tag{6.28}$$

The Q-tuning system of Figure 6.37 is an amplitude locked loop which operates using this proportionality between gain and Q. It is assumed that separate frequency-tuning circuits maintain ω_0 of the filter exactly equal to the desired value, and that the gain of the filter is equal to the Q at ω_0. The reference signal is attenuated by a factor $1/K_Q$ by a potential divider, and applied to the filter input. The output amplitude of the filter is therefore $V_{ref}Q/K_Q$. A pair of matched envelope detectors generate DC levels proportional to V_{ref} and the filter output, which are compared by an error amplifier. The resulting feedback signal varies the Q of the filter so that the filter output is equal to V_{ref}, in which case $Q = K_Q$. Since K_Q is determined by component ratios which can be made accurately, Q is also accurately defined.

6.4.6   Tuning of high-order leap-frog filters

Multiple loop feedback filters are desirable for fully integrated filters because of their low sensitivity. This applies especially to high-Q bandpass filters, due to their high sensitivity to frequency tuning errors, and the high Q required from each filter section. However, as pointed out in Section 6.3, the multiple feedback structure which is responsible for this low sensitivity at the same time makes this type of filter more difficult to tune. One method of overcoming this difficulty is to use master–slave tuning systems; however in the case of high-Q bandpass filters, it may not be possible to ensure sufficiently accurate matching between sections of the filter circuit to obtain adequate tuning accuracy. This section describes a tuning method applicable to the leapfrog (LF) form of MLF filter, and other types of filter based on LC ladder simulation, in which the individual pole frequencies are tuned without relying on precise matching [38].

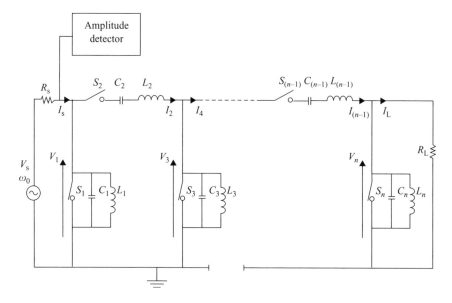

Figure 6.38   LC bandpass tuning using Dishal's method

The same tuning problem exists when tuning passive LC bandpass filters, such as the ladder filter shown in Figure 6.38. Synthesis of this ladder filter with centre frequency ω_0 results in the inductor and capacitor values in each branch of the ladder having the same resonant frequency, $1/L_iC_i = \omega_0^2$. One method of tuning the filter would be to isolate each ladder branch, resulting in several separate second-order bandpass sections, which could then be individually tuned to the correct centre frequency. The drawback of this approach is that considerable additional circuitry would be required to route the test signal to and from each resonator in turn. This routing circuitry would also introduce additional parasitic capacitance and cross-talk between filter sections, degrading filter performance. An alternative method of tuning this filter which avoids some of these problems is due to M. Dishal [39]. This has been widely applied to many forms of passive bandpass filter. It has the advantage of requiring no additional signal paths connected to the internal nodes of the filter.

To tune the filter of Figure 6.38, initially all switches in the series arms are open, and all those in the shunt arms are closed. A signal is applied to the input at frequency ω_0, and V_1 is monitored by the amplitude detector. S_1 is opened, and C_1/L_1 are tuned to parallel resonance, i.e. maximum amplitude of V_1. Since S_2 is open, the resonator C_1/L_1 is isolated from the rest of the circuit, which therefore does not alter the resonant frequency. Next, S_2 is closed, and C_2/L_2 are tuned to series resonance and minimum V_1. Since S_3 is closed, C_2/L_2 are also isolated from succeeding stages of the filter. Each successive branch is then adjusted in turn, the shunt branches for maximum V_1 and the series branches for minimum V_1, with the associated switch being opened or closed. Since all preceding branches are already resonant, the reactive component of their net series or shunt impedance is zero, and they are 'transparent' at frequency ω_0. When L_n/C_n have been adjusted, the tuning process is complete. In tuning schemes for second-order cascade filters, it is normally necessary to provide Q-tuning capability. This is not done when tuning using Dishal's method as described, and so the tuning process does not completely define the transfer function of the filter. The bandwidth and ripple in the response are defined by ratios between component values in different branches of the circuit, whilst the method described above only tunes the inductor and capacitor in each individual branch in isolation. However, because all branches are resonant at ω_0, the passband is symmetrical, insertion loss is minimised, and gross distortion of the frequency response does not occur.

As discussed in Section 6.3, a widely used technique for active filter design is to simulate the function of passive LC ladder filters using active elements. Two methods for doing this are the component substitution technique, and the leap-frog MLF filter. Clearly, using the component substitution technique, the inductors of Figure 6.38 could be replaced by gyrators which performed the same circuit function, and Dishal's method could still be applied to tune the circuit. A leap-frog MLF filter which simulates the filter of Figure 6.38 is shown in Figure 6.39. In this structure, the reactive passive components are replaced by integrators.

Examination of this block diagram shows that each LC resonator in the prototype is replaced by a two-integrator loop bandpass biquad of the type shown in Figure 6.5, with the same resonant frequency, ω_0, and transfer function of the isolated LC circuit. As in the case of the LC bandpass filter, it would be possible to tune each resonator separately, as is proposed in Reference 40, but with the same drawbacks with regard to signal routing. In the LC filter, coupling between resonators occurs because of their ladder connection; in the LF filter, this coupling occurs via the feedback paths. The switches in the feedback paths of Figure 6.39 perform an equivalent function as in Figure 6.38, isolating the resonators from one another. Dishal's method can therefore

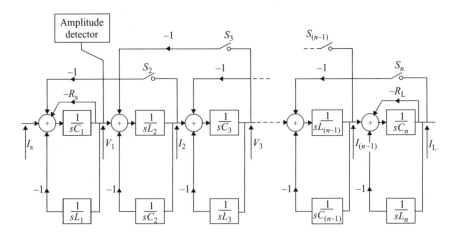

Figure 6.39   LF simulation of LC bandpass filter

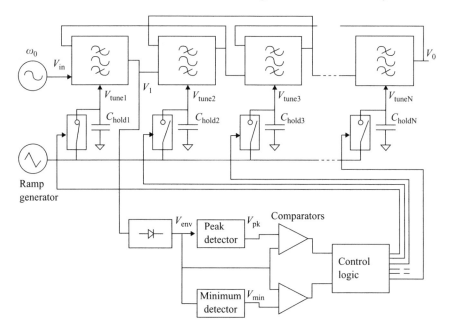

Figure 6.40   Peak-detecting tuning scheme extended to LF bandpass filter

be applied to the MLF filter, tuning the resonant frequency of each biquad in turn for alternate maximum or minimum amplitude at V_1, the output of the first integrator. V_1 is analogous to the voltage at the input termination of the ladder filter prototype.

An on-chip tuning system which tunes the pole frequency of a single biquad by detecting the peak of its amplitude response is described in Section 6.4.4. This scheme may be extended as in Figure 6.40 to sequentially tune a number of biquads making up the bandpass LF filter. Initially, V_{tune1} is adjusted for peak output at V_1. To isolate the first biquad from the rest of the filter, $V_{tune2} \ldots, V_{tuneN}$ are initialised to zero, debiasing the other biquads. After V_{tune1} has been adjusted, V_{tune2} is tuned for minimum V_1. The minimum detector is a peak detector with inverted polarity. The process is repeated with $V_{tune3}, \ldots, V_{tuneN}$ until all biquads have been tuned.

6.5   Conclusions

We have examined some of the most widely used techniques for designing fully integrated continuous-time filters, and, of equal importance, methods by which these filters can be automatically tuned by on-chip circuitry. The frequency range over which these filters may be applied in wireless transceivers is extremely wide, ranging from audio frequencies to the RF signal frequency, which is normally in the UHF range, and with the general availability of deep-submicron IC processes is being extended to several gigahertz.

Integrated filter architectures can be divided into second-order cascade and multiple loop feedback types; a subset of the MLF filters includes a range of passive LC ladder simulation techniques. The second-order cascade offers simplicity of design with a modular circuit structure; however it possesses relatively high sensitivity compared to the MLF structures. Filters based on simulations of LC ladders are probably the most popular for integrated filter design. Their low sensitivity helps to offset the difficulties caused by loose integrated component tolerances. These may utilise component substitution and circuit transformation techniques to eliminate inductors from the prototype ladder design, or be based on 'leap-frog' structures which simulate ladder filter behaviour at a system level.

The OTA-C filter has important benefits for high-frequency operation. In this structure, all high-impedance nodes have grounded capacitors connected to them to define the filter response. This allows the inevitable stray capacitances to be absorbed into these tuning capacitors, without introducing additional unwanted parasitic poles into the response. The operational transconductance amplifiers operate in an open-loop mode; relatively low open-loop gain is required, which can be achieved by amplifier structures with wide bandwidth and low power consumption. The OTA is also well suited to electrical tuning of its transconductance, OTA-C filters have been successfully used for frequencies ranging from audio to several hundred megahertz. The main difficulties in designing OTA-C filters lie in the non-ideal behaviour of the OTAs used. The transistors making up the OTA are inherently non-linear, limiting the dynamic range of the filter. The finite output impedance of the transistors has the effect of reducing the maximum filter Q which can be achieved. Linearity and output impedance can be improved by resorting to more complex circuits; however these tend to have reduced bandwidths.

The chief advantage of the MOSFET-C technique is that good linearity can be readily achieved due to the high loop gain within the integrator circuits, and the distortion-cancelling properties of the balanced MOSFET-R structure. MOSFET-C filters have been very widely used as IF filters in receivers because of their low distortion capability. Typically, they are used from low frequency to tens of megahertz; the upper frequency limiting factor is the need for the amplifier gain–bandwidth product to be large in comparison with the operating frequency in order to achieve low distortion, and to suppress the effects of parasitic capacitance at the amplifier input and output. This usually means that, using a particular IC technology, higher operating frequency can be achieved using OTA-C techniques. However, with the very large bandwidths now achieved by submicron CMOS processes, this may not be a limiting factor, and successful MOSFET-C designs have been implemented for frequencies exceeding 100 MHz [21].

Active-LC filters are a recent development in integrated IC filter techniques. They promise some of the advantages of passive LC filters since the role of the active components in the filter is simply to replace energy dissipated in the losses of the passive components. Thus, potential advantages are low power consumption and good linearity at the highest frequencies. Several difficulties remain: the low Q and restricted range of values feasible for integrated spiral inductors in current IC processes, the difficulty of providing accurate compensation of losses over a wide frequency range and the difficulty of providing frequency and Q tuning. Nonetheless, active-LC filters with useful performance at frequencies of a few GHz have been described.

Tuning remains a critical issue for integrated filter designs. Virtually all filters employ some form of frequency tuning. With careful design, including the accurate modelling of parasitic effects, many designs operating at moderate frequencies (in the megahertz range), and with moderate Q (filter bandwidth of the same order as the centre frequency) have successfully achieved sufficient accuracy using relatively simple master–slave frequency tuning schemes without the need for Q tuning. For high-frequency and high-Q designs, Q tuning is required, due to the high sensitivity of Q to parasitic effects in the circuit. For such filters, the accuracy of matching between sections of the filter circuit may not be adequate to allow the use of master–slave techniques, so design of the tuning system may be much more difficult than the filter itself. Therefore, the filter design may not be feasible because provision of on-chip tuning is too difficult. One reason for the popularity of the low-IF superhet receiver architecture is the absence of requirements for narrowband filters. Improved tuning techniques for high-order filters operating at high frequency and Q filters remain an important research topic.

The field of integrated continuous-time filters is constantly evolving, being driven by progress in IC fabrication techniques. In the immediate future, the trend towards smaller geometry CMOS processes primarily designed for digital applications presents the filter designer with new problems. For example, reduced supply voltages limit the signal amplitudes that can be processed linearly within the filter circuit, without corresponding reduction in noise level. Thus, it is difficult to maintain dynamic range. MOSFETs with very short channel lengths exhibit low output resistance, making OTA design more difficult, especially since many circuit techniques for improving OTA performance are made impractical by the low supply voltage. There are also positive aspects; smaller geometries result in wider amplifier bandwidths, while copper interconnects and low K dielectrics should result in reduced parasitics and higher performance integrated inductors. Looking further ahead, new IC processes, using Si-Ge, or different semiconductor materials may become important, and require the development of new circuit techniques. Techniques such as MEMS and nanotechnology are already of commercial interest; if they become a feature of mainstream IC processes, a whole new range of components may become available to the integrated filter designer.

6.6   References

1 'BlueCore2-external production information data sheet', Cambridge Silicon Radio, Cambridge, 2002
2 DURDODT, C., FRIEDRICH, M., GREWING, C., HAMMES, M., HANKE, A., HEINEN, S., OEHM, J., PHAM-STABNER, D., SEIPPEL, D., THEIL, D., VAN WAASEN, S., and WAGNER, E.: 'The first very low-IF RX, 2-point modulation TX CMOS system on chip Bluetooth solution'. Proceedings of 2001 IEEE Radio Frequency Integrated Circuits Symposium, Arizona, pp. 99–102

3 RAZAVI, B.: 'Challenges in portable RF transceiver design', *IEEE Circuits and Devices Magazine*, 1996, **12** (5), pp. 12–25
4 HUGHES, J. B., SPENCER, A., WORAPISHET, A. and SITDHIKORN, R.: '1 mW CMOS polyphase channel filter for Bluetooth'. IEE Proceedings on *Circuits, Devices and Systems*, October/December 2002, **149** (5/6), pp. 348–54
5 NIMMO, R.: 'Analogue electronics, the poor relation?'. Proceedings IEE Symposium on Analogue Signal Processing, Oxford, 1 November 2000, pp. 1/1–1/5
6 ZVEREV, A. I.: 'Handbook of filter synthesis' (John Wiley, New York, 1967)
7 LUTOVAC, M. D., DEJAN, V. T. and EVANS, B. L.: 'Filter design for signal processing' (Prentice-Hall, New Jersey, 2001)
8 DELIYANNIS, T., SUN, Y. and FIDLER, J. K.: 'Continuous-time active filter design' (CRC Press, Florida, USA, 1999)
9 SU, K. L.: 'Analogue filters' (Chapman and Hall, London, 1996)
10 SCHAUMANN, R. and VAN VALKENBURG, M. E.: 'Design of analogue filters' (Oxford University Press, New York, 2001)
11 SUN, Y. and FIDLER, J.K.: 'Structure generation and design of multiple loop feedback OTA-grounded capacitor filters', *IEEE Transactions on Circuits and Systems – I*, January 1997, **44** (1), pp. 1–11
12 MARTIN, K. and SEDRA, A. S.: 'Design of signal-flow graph (SFG) active filters', *IEEE Transactions on Circuits and Systems*, April 1978, CAS-**25**, pp. 185–95
13 SCHAUMANN, R.: 'Simulating lossless ladders with transconductance-C circuits', *IEEE Transactions on Circuits and Systems – II*, March 1998, **45** (3), pp. 407–10
14 SU, H. W. and SUN, Y.: 'A CMOS 100 MHz continuous-time seventh order 0.05° equiripple linear phase multiple loop feedback G_m-C filter'. Proceedings of IEEE ISCAS 2002. Arizona, II, pp. 17–20
15 SUN, Y.: 'Design of high frequency integrated analogue filters' (The Institution of Electrical Engineers, London, 2002)
16 SANCHEZ-SINENCIO, E. and SILVA-MARTINEZ, J.: 'CMOS transconductance amplifiers, architectures and active filters: a tutorial', *IEE Proceedings: Circuits, Devices and Systems*, February 2000, **147** (1), pp. 3–13
17 GLINIANOWICZ, J., JAKUSZ, J., SZCZEPANSKI, S. and SUN, Y.: 'High frequency two-input CMOS OTA for continuous-time filter applications'. IEE Proceedings: *Circuits Devices and Systems*, February 2000, **147** (1), pp. 13–18
18 SILVA-MARTINEZ, J., STEYAERT, M.S.J. and SANSEN, W.M.C.: 'A large-signal very low-distortion transconductor for high-frequency continuous-time filters', *IEEE Journal of Solid-State Circuits*, July 1991, **26** (7), pp. 946–55
19 TANIMOTO, H., KOYAMA, M. and YOSHIDA, Y.: 'Realization of a 1 V active filter using a linearization technique employing a plurality of emitter-coupled pairs', *IEEE Journal of Solid-State Circuits*, July 1991, **26** (7), pp. 937–45
20 PANKIEWICZ, B., WOJCIKOWSKI, M., SZCZEPANSKI, S. and SUN, Y.: 'A field programmable analogue array for CMOS continuous-time OTA-C filter applications', *IEEE Journal of Solid-State Circuits*, February 2002, **37** (2), pp. 125–36

21 GROENEWOLD, G.: 'Low power MOSFET-C 120 MHz Bessel allpass filter with extended tuning range', *IEE Proceedings: Circuits, Devices and Systems*, February 2000, **147** (1), pp. 28–34

22 BANU, M. and TSIVIDIS, Y.: 'Fully active integrated RC filters in MOS technology', *IEEE Journal of Solid-State Circuits*, December 1983, SC-**18** (6), pp. 644–51

23 CZARNUL, Z.: 'Novel MOS resistive circuit for synthesis of fully integrated continuous-time filters', *IEEE Transactions on Circuits and Systems*, July 1986, **33** (7), pp. 718–21

24 CHENG, Y., GONG, J. and WU, C. Y.: 'New CMOS 2 V low-power IF fully differential Rm-C bandpass amplifier for RF wireless transceivers', *IEE Proceedings: Circuits, Devices and Systems*, December 2001, **148** (6), pp. 318–22

25 TANG, C. C., WU, C. H. and LIU, S. I.: 'Minature 3-D inductors in standard CMOS process', *IEEE Journal of Solid-State Circuits*, April 2002, **37** (4), pp. 471–80

26 LI, D. and TSIVIDIS, Y.: 'Active LC filters on silicon', *IEE Proceedings: Circuits Devices and Systems*, February 2000, **147** (1), pp. 49–56

27 SOORAPANTH, T. and WONG, S. S.: 'A 0-dB IL 2140 ± 30 MHz bandpass filter utilising Q-enhanced spiral inductors in standard CMOS', *IEEE Journal of Solid-State Circuits*, May 2002, **37** (5), pp. 579–86

28 KUHN, W. B., YANDURU, N. K. and WYSZYNSKI, A. S.: 'Q-enhanced LC bandpass filters for integrated wireless applications', *IEEE Transactions on Microwave Theory and Techniques*, December 1998, **46** (12), pp. 2577–85

29 LI, D. and TSIVIDIS, Y.: 'Design techniques for automatically tuned integrated gigahertz-range active LC filters', *IEEE Journal of Solid-State Circuits*, August 2002, **37** (8), pp. 967–77

30 KUHN, W. B., ELSHABINI-RIAD, A. and STEPHENSON, F. W.: 'A new tuning technique for implementing very high Q, continuous-time, bandpass filters in radio receiver applications', *Proceedings of IEEE ISCAS 1994*, London, **5**, pp. 257–60

31 KRUMMENACHER, F. and JOEHL, N.: 'A 4 MHz CMOS continuous-time filter with on-chip automatic tuning', *IEEE Journal of Solid-State Circuits*, June 1988, **23** (3), pp. 750–8

32 SHI, B., SHAN, W. and ANDREANI, P.: 'A 57 dB image band rejection CMOS G_m-C polyphase filter with automatic frequency tuning for Bluetooth', *Proceedings of IEEE ISCAS 2002*, Arizona, **V**, pp. 169–72

33 YAMAZAKI, H., OISHI, K. and GOTOH, K.: 'An accurate center frequency tuning scheme for 450 kHz CMOS G_m – C bandpass filters', *IEEE Journal of Solid-State Circuits*, December 1999, **34** (12), pp. 1691–7

34 PHAM, T. K. and ALLEN, P. E.: 'A design of a low-power, high accuracy, and constant-Q-tuning continuous-time bandpass filter', *Proceedings of IEEE ISCAS 2002*, Arizona, **IV**, pp. 639–42

35 KARSILAYAN, A. I. and SCHAUMANN, R.: 'Mixed-mode automatic tuning scheme for high-Q continuous-time filters', *IEE Proceedings: Circuits, Devices and Systems*, February 2000, **147** (1), pp. 57–64

36 LINARES-BARRANCO, B. and SERRANO-GOTARREDONA, T.: 'A loss control feedback loop for VCO stable amplitude tuning of RF integrated filters', *Proceedings of IEEE ISCAS 2002*, **I**, pp. 521–4

37 LI, D. and TSIVIDIS, Y.: 'Design techniques for automatically tuned gigahertz range active LC filters', *IEEE Journal of Solid-State Circuits*, August 2002, **37** (8), pp. 967–77

38 MORITZ, J. R. and SUN, Y.: 'Automatic tuning of high frequency, high Q multiple loop feedback bandpass filters', *Proceedings of IEEE ISCAS 2002*, Arizona, **V**, pp. 605–8

39 DISHAL, M.: 'Alignment and adjustment of synchronously tuned multiple resonant circuit filters', Electrical Communication, June 1952, pp. 154–64

40 BLISS, W. G. and SADKOWSKI, R.: 'In situ tuning monolithic continuous time leapfrog filters', *Proceedings of the IEEE 35th Midwest Symposium on Circuits and Systems*, Washington DC, August 1992, **2**, pp. 938–41

Chapter 7

Low-voltage integrated RF CMOS modules and frontend for 5 GHz and beyond

Mourad N. El-Gamal and Tommy K. Tsang

7.1 Introduction

The growing number of users and the demand for high-speed wireless communications has motivated designers to move from the 1–2 GHz range towards higher frequency bands. Recently, new standards in the 5 GHz range for wireless local area network (WLAN) applications have been defined, such as the IEEE 802.11a standard for the FCC unlicensed national information infrastructure (U-NII) band in the US, and the high performance radio LAN (HIPERLAN) standard in Europe.

Traditionally, radio frequency integrated circuits (RFICs) were implemented in GaAs or SiGe bipolar technologies, because of their relatively high unity gain cut-off frequencies f_T (i.e. >65 GHz) and their superior noise performance. However, as the minimum feature size of CMOS devices decreases, the f_T of the transistors continues to improve to the point where it is becoming comparable to those of GaAs and SiGe processes. Deep-submicron CMOS devices with f_T's exceeding 100 GHz and minimum noise figures (NF) less that 0.5 dB at 2 GHz have been demonstrated [1]. Because of these promising RF performances, together with the advantages of low cost and ease of integration with baseband digital circuitry, CMOS is becoming a viable alternative for RF applications, with continuous efforts towards implementing higher frequency circuitry operating from lower supply voltages (e.g. [2–15]).

This chapter starts by introducing a CMOS low noise amplifier (LNA) architecture suitable for operation from sub-1 V supplies. Experimental results from several fabricated prototype chips are presented to demonstrate the versatility of this topology. Namely, the first prototype operates at a centre frequency of 5.8 GHz, with gain and frequency tuning capabilities. The second and third prototypes were optimised for future RF applications, operating at higher carrier frequencies of 8–9 GHz, also featuring gain control. Implemented in a standard 0.18 μm CMOS process, all circuits exhibit a power gain of greater than 12 dB with a noise figure as low as 2.5 dB, when operating from a 1 V power supply. All implementations have a gain tuning range of over 10 dB, and can operate from a supply voltage as low as 0.7 V. The LNAs use no off-chip components, allowing for a simple and robust integration.

The chapter then proceeds with introducing a new LC-based oscillator structure, which also enables operation from very-low supply voltages (0.85 V), while being suitable for high-frequency RF applications. Two 0.18 µm CMOS VCO prototypes are reported. The 8.7 GHz VCO operates from a supply voltage of 0.85 V, consumes 6 mW, and exhibits -100 dBc/Hz phase noise at a 600 kHz offset. The 10 GHz prototype operates from a supply voltage of 1 V, consumes 9 mW, and has -98 dBc/Hz phase noise at a 600 kHz offset. A tuning range of 400–450 MHz is achieved without using varactors.

Finally, in order to demonstrate the potential of using CMOS for low-voltage, high-frequency applications, the chapter concludes with the presentation of a 5 GHz receiver frontend operating from a 0.8 V supply.

7.2   A versatile low-voltage CMOS LNA topology

The low noise amplifier is a critical building block in any RF receiver. It is usually the first active circuit in the received signal path. Its role is to amplify the weak input RF signal, while introducing as little noise as possible. Low noise figure, high linearity, enough gain, and low power consumption are the main performance merits of an LNA design.

With the targeted voltage supply down to 1 V, there is a limited number of suitable LNA topologies available to designers. Amplifier architectures which can operate from low voltage supplies are shown in Figure 7.1. Although the discussion in this chapter is mainly focused on the CMOS technology, some of the low-voltage topologies presented can be generalised to other technologies, such as silicon bipolar [16].

7.2.1   Single-transistor architecture

Consider the single-transistor amplifier shown in Figure 7.1a. Although the supply voltage can be made as low as the saturation voltage of the CMOS transistor, i.e. V_{sat}, the practical minimum supply voltage needed will have to be higher than the threshold voltage of the transistor necessary to properly bias it and to turn it on ($V_{gs} \gg V_{th}$). In spite of meeting the very low supply requirement, there are many practical challenges with this circuit. Most importantly, this topology is very susceptible to instability problems, especially at RF. This is mainly due to the gate-to-drain parasitic capacitance (C_{gd}) which provides both forward and reverse low impedance paths between the input and output signals. This considerably complicates impedance and noise matching, since the input reactance is no longer dictated only by the gate-to-source capacitance (C_{gs}), but is also dependent on a correlated function of the voltage gain (A_v) and C_{gd}. The equivalent input capacitance of the transistor (C_{in}) is

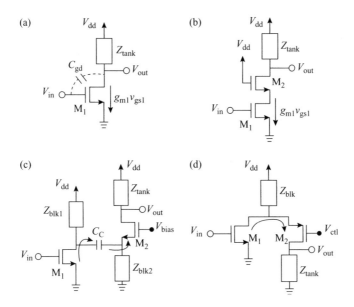

Figure 7.1 LNA topologies: (a) single transistor, (b) conventional cascode, (c) LC-coupled, (d) modified folded cascode

described by

$$C_{in} = C_{gs} + (1 + A_v)C_{gd}. \tag{7.1}$$

It should also be noted that this single-transistor structure suffers from the Miller effect, which limits its bandwidth.

7.2.2 Conventional cascode LNA architecture

In order to alleviate the limitations of the single-transistor LNA, it is very common to use the cascode configuration shown in Figure 7.1b. The main limitation of this structure is the need for a relatively high supply voltage headroom, since it involves the stacking of at least two transistors. This is not quite suitable for very low voltage applications. As the minimum feature sizes of the transistors are shrinking, the threshold voltage (V_{th}) and the drain-to-source voltage (V_{ds}) are decreasing. It is therefore becoming more and more feasible to implement a low-voltage (i.e. <1 V) cascode LNA in state-of-the-art deep-submicron technologies. However, the transistors will probably be forced to operate close to their triode mode, resulting in a serious degradation in linearity.

7.2.3 LC-coupled architecture

The need to improve the linearity of the conventional cascode structure, while allowing operation from a very low supply voltage, has motivated the development of the

LC-coupled LNA topology shown in Figure 7.1c. The main idea behind this topology is to decouple the AC and DC currents in the two transistors, hence allowing the reduction of the voltage supply, without the need to push the transistors to operate close to triode. To ensure that the circuit operates as a cascode amplifier, the entire RF signal current ($g_{m1} v_{gs1}$) generated by M_1 should be fed into the source of M_2 (i.e. driving $1/g_{m2}$). For this to be achieved, two conditions need to be met simultaneously. First, the impedance (Z_{blk1}) at the drain of M_1 should be much larger than the impedance looking towards the coupling capacitor and the source of M_2. Second, the impedance (Z_{blk2}) at the source of M_2 should be much larger than the impedance looking into the source of M_2. These conditions can be summarized as follows

$$Z_{tank1} \gg \frac{1}{j\omega C_C} + \left[Z_{tank2} \| \frac{1}{g_{m2}} \right],$$ (7.2)

$$Z_{tank2} \gg \frac{1}{g_{m2}}.$$ (7.3)

In essence, this can be achieved by using inductors as large as possible for the blocking impedances, and a coupling capacitor that is as large as possible. Since the exact value of the blocking inductor is not critical, bonding wires could be used, provided that the bonding pad capacitive losses are accounted for, and that the inductance of the shortest possible bonding wire (this can be determined from the physical dimensions of the die, and the size of the cavity of the package used) will still meet the conditions in equations (7.2) and (7.3). If that is not the case, then on-chip inductors could be used in series with the bonding wire inductors, in order to ensure that the minimum blocking impedance needed is obtained. This will, however, result in extra losses, can considerably complicate the design, and will result in a larger die area, thus higher cost. The need for large coupling capacitors is a major disadvantage of this structure. In addition to the increase in area and cost, large integrated capacitors suffer from high RF signal losses to the substrate. A design compromise will therefore be needed regarding the size of the capacitor to be used, which probably will result in allowing finite signal loss across the capacitor, in order to minimise its size, at the price of a finite loss in gain.

7.2.4   Folded cascode architecture

Motivated by the limitations of the LC-coupled architecture, the folded cascode LNA topology, shown in Figure 7.1d, is adopted in this work. It does not require the use of large coupling capacitors. Historically, the use of PMOS devices in RF circuits was not common due to their lower f_Ts, compared to their NMOS counterparts. As CMOS technologies scale down to 0.18 μm and beyond, the f_Ts of the PMOS devices are becoming in the order of 20 GHz, making them a good candidate for high-performance RF designs. An added advantage of the use of PMOS transistors is their lower noise [17].

The structure in Figure 7.1d is borrowed from the conventional folded cascode topology of CMOS operational amplifiers, with modifications making it further suitable for low-voltage RF LNA applications. For the circuit in Figure 7.2a, the NMOS

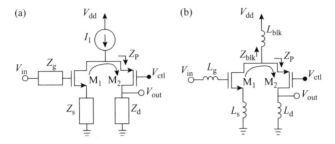

Figure 7.2 *Folded cascode topologies: (a) wideband conventional, (b) narrowband modified*

transistor M_1 amplifies the input signal, while the PMOS transistor M_2 acts as a current buffer. The current source (I_1) exhibits a wideband small-signal, high-impedance response. Thus, all the signal current ($g_{m1} v_{gs1}$) generated by M_1 is forced into the source of M_2. In RF applications, it is often desired to implement the LNA to have a frequency selective response, in order to minimise out-of-band noise and interference. Due to the very high frequency of operation (in the gigahertz range), it is feasible to implement a frequency selective circuit using inductors and capacitors of reasonably small sizes. This is done by replacing the current source (I_1) by a narrowband LC resonant tank, as shown in Figure 7.2b. Similar to the case of the LC-coupled topology, it is possible to use high-quality bonding wires for the inductor. This is done at no extra cost, since an inductor would anyway be used to connect the LNA to an off-chip power supply. For large values of inductances, additional inductive traces on the printed circuit board could be used. A total inductance value of 5 nH can easily be achieved, which would present a relatively high impedance to the RF signal. The blocking condition is given by

$$Z_{blk} \approx j\omega_o L_{blk} \gg Z_P, \tag{7.4}$$

where $Z_P \approx (1/g_{m2}) \| r_{o2}$, and g_{m2} and r_{o2} are the transconductance and output resistance of transistor M_2 respectively.

7.2.5 Gain control in the folded cascode structure

Linearity is becoming an important system design issue in today's wireless applications. Narrower allocated channel spacing drives the need for high linearity (i.e. IIP3), in addition to good noise and gain performance. Since high linearity usually trades off with high gain characteristics, gain controllability becomes a very desirable feature in modern LNA designs. This can enhance the IIP3 by decreasing the gain at high input-power levels. It also relaxes both the linearity requirements of other blocks in the receiver chain, and the dynamic ranges of the on-chip variable gain amplifiers (VGAs) at later stages.

There exist several types of variable gain amplifier (VGA) solutions in the literature (e.g. [18, 19]) as shown in Figure 7.3. They include (i) a switch-control type,

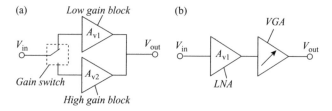

*Figure 7.3   Conceptual views of variable gain amplifiers: (a) switch-control type,
(b) two-stage LNA–VGA type*

which provides gain control by switching on/off active gain components, and (ii) the
two-stage LNA–VGA type, which achieves gain control through the use of a VGA
as a second stage. This extra gain-control functionality comes at the price of higher
circuit complexity, which translates into an increase in power consumption and noise
degradation.

One of the main features of the topology in Figure 7.2b is that it readily provides
gain control, without any extra circuit complexity. Gain control can be simply achieved
by varying the gate voltage (V_{ctl}) of M_2, which varies the impedance (Z_P) looking
into the source of the transistor, resulting in an overall variation of the gain of the LNA.
We define the gain-tuning factor (G_{tune}) as representing the portion of the AC signal
current, generated by the input transistor M_1, which flows into the source of M_2. It is
given by

$$G_{tune} = \frac{Z_{blk}}{Z_{blk} + Z_P} = \frac{Z_{blk}(1 + g_{m2}r_{o2})}{Z_{blk}(1 + g_{m2}r_{o2}) + r_{o2}}. \tag{7.5}$$

It is evident that G_{tune} can be adjusted by varying g_{m2}, which is controlled by V_{ctl}.
Thus, the voltage gain (A_{tune}) of the folded cascode structure in Figure 7.2b can be
shown to be

$$A_{tune} = A_v G_{tune} = g_{m1}\frac{Q_{in}}{2}Z_{d2}G_{tune}, \tag{7.6}$$

where A_v is the fixed voltage gain of a conventional cascode LNA [20], Z_{d2} is the
impedance level at the drain of M_2, and Q_{in} is the quality factor of the input series
RLC resonant tank.

Note that gain control is achieved without affecting the input noise and impedance
matching, which are set independently by transistor M_1. Hence, a low noise figure
can be achieved while providing gain tuning. This is a considerable advantage when
compared to other LNA architectures. For example, conventional cascode amplifiers
(Figure 7.1b) do not have this flexibility in gain-control. The overall gain of the LNA
is governed by transistor M_1, while the cascode transistor M_2 only acts as a current
buffer. Gain control could only be achieved by altering the biasing current of M_1,
directly affecting the input noise and impedance matching, which is definitely not
desirable.

To conclude, the modified folded cascode topology is a promising low-voltage structure, well-balanced in terms of stability and noise, with an inherent advantage of simple gain control compared to other architectures. In this work, all the LNA designs presented are based on this architecture, with a targeted operating frequency of 5 GHz and beyond.

7.2.6 Frequency control in the folded cascode structure

Frequency tunability can also be a desirable feature in LNA designs, as it can serve two main purposes: (i) to compensate for process variations and inaccuracies in inductor modelling, and (ii) to tune for a different receive band centre frequency. Frequency tuning was explored by adding a varactor at the resonant tank of the LNA as shown in Figure 7.4. There are a number of ways to implement a varactor. The one used in this prototype is an NMOS transistor over an Nwell, which makes use of the gate capacitance of the transistor [21, 22]. This structure has been reported to provide a wider tuning range and a better quality factor Q, in comparison to other PN or P$^+$N junction-based varactors. This ensures a high Q for the resonant tank. For the prototypes with frequency tuning presented in this work, the varactors were designed to achieve a 10 per cent frequency tuning range.

7.2.7 Design equations and optimisation

Designing CMOS RF circuits operating at frequencies greater than 1 GHz imposes many challenges and difficulties. Iterative circuit optimisation is often necessary. In this section, design equations, as well as a summary of inductor design guidelines and layout techniques, which have been adopted in this work, are presented.

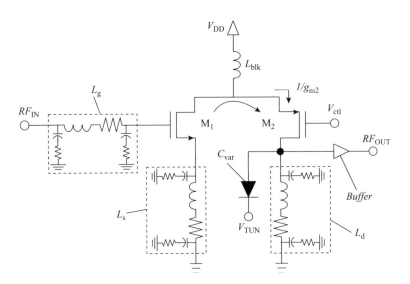

Figure 7.4   A single-ended fully tuneable (gain and frequency) low-voltage LNA

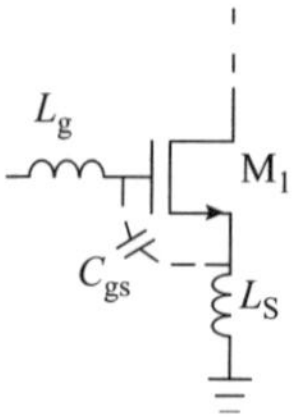

$$Z_{in} = s(L_S + L_g) + \frac{1}{sC_{gs}} + \left(\frac{g_{m1}}{C_{gs}}\right)L_S$$

$$\approx \left(\frac{g_{m1}}{C_{gs}}\right)L_S = \omega_T L_S \text{ (at resonance)},$$

$$\omega_0 = \frac{1}{\sqrt{(L_g + L_S)C_{gs}}}.$$

Figure 7.5   Conditions for input impedance matching

7.2.7.1   Input impedance matching

Input matching is an important design issue, necessary to minimise signal reflection and noise. There is often a trade-off between noise and input impedance matching in LNA designs. This trade-off reflects on the choice of transistor sizing, which is mainly dependent on the designer's objectives and priorities. A common approach used is to first determine the transistor sizing which makes the circuit be approximately noise matched to the characteristic impedance of the system, typically 50 Ω, at the frequency of interest. Then, a minimal passive network is added to fine tune the input matching. For the LNAs in this chapter, a series-connected two-element matching network was used, as shown in Figure 7.5. It consists of gate and source inductors L_g and L_S, respectively. The source degeneration inductor L_S is used to match the real part of the input impedance to the characteristic impedance ($Z_s = 50 \Omega$), while the combination of source and gate inductors is used to cancel out the reactance due to the parasitic capacitance C_{gs} of the input transistor M_1. The conditions for input impedance matching (Z_{in}), and the expression for the resonant frequency ω_0 are summarized as shown in Figure 7.5. Since a prime objective is noise minimisation, the matching network is designed to achieve the minimum overall noise figure, while still maintaining a reasonable input impedance matching, at the frequency of interest.

7.2.7.2   Integrated inductor design guidelines in CMOS

Proper modelling of integrated inductors at radio frequencies is one of the most challenging and crucial tasks in designing successful RFICs. Standard CMOS integrated inductors have inherently low quality factors (i.e. $Q < 5$) since they exhibit serious substrate and dielectric losses, which become dominant at GHz frequencies. This is mainly due to the high conductivity of the CMOS substrate. Substantial research efforts have been invested into this topic recently, attempting to improve the accuracy of inductor modelling and the quality factor of inductors in a CMOS technology. The main objective of this section is to provide guidelines and point out the trade-offs involved in inductor design in CMOS at RF.

Based on earlier work from the literature on inductor design (e.g. [23, 24]), there are a number of approaches to implement integrated inductors. Apart from the key geometrical parameters such as conductor width (W), number of turns (N), and inductor shape, the use of patterned ground shields and multilayer structures are other inductor design issues. Since the targeted operating frequencies in this work are at or above 5 GHz, the prime objective is to design inductors with high quality factors (i.e. $Q > 5$) at those frequencies, and to simultaneously maximise their self-resonance frequencies, f_{RES}, pushing them much higher than this range.

Although the use of a patterned ground shield will reduce eddie current losses at high frequency, and improve the quality factor, it can significantly reduce the f_{RES} of an inductor due to the additional capacitive parasitics it introduces. Hence, in this work, a patterned ground shield is not inserted between the spiral inductor and the silicon substrate. Furthermore, the use of multilayer inductor structures connected in series in order to increase the self-inductance per unit area is not necessary, since the required inductances are in the order of 1–1.6 nH, which can be easily implemented with one simple planar structure. In order to achieve high Q through reducing the series resistance, the number of turns (N) should be minimised, while the conductor width (W) should be made large. However, increasing W can have a negative effect on the self-resonance frequency, since wider metal traces translate into a larger parasitic capacitance to substrate. There exists an optimum value for W, which we found to be 20 μm. This result was obtained and verified by both simulation and experimental data [25]. Finally, to further reduce the series resistance of the inductor, two or three metal layers connected in parallel are used to emulate a thicker conductor.

7.2.7.3 RF layout techniques

Apart from proper modelling of integrated passives, layout is another important step in optimising high-frequency designs. Poor layout could result in large discrepancies between the actual and expected performance, or even result in a non-working circuit. The layout of a 5.8 GHz CMOS LNA is shown in Figure 7.6, to illustrate the RF layout techniques used in this work.

Careful layout is observed in order to maximise performance. The layout is done in a uni-directional fashion, i.e. no signal returns close to its origin, to avoid coupling back to the input. The RF input and output ports are placed on opposite sides of the chip to improve port-to-port isolation. Since on-chip probing is used to measure the LNAs performance, standard ground-signal-ground (GSG) pad configurations are used at both the input and output RF ports. Furthermore, the signal pads (i.e. RF input and output) are implemented using the top metal layer only, in order to reduce RF signal loss through parasitic capacitances to the substrate. In order to minimise the effect of substrate noise on the system, a solid ground plane, constructed using a low resistivity metal-1 material, is placed between the signal pads (metal-6) and the substrate.

Since the operation of inductors involves magnetic fields, they can affect nearby signals and circuits, and cause interference. Therefore, inductors are placed far apart from each other, as well as from the main circuit components, with reasonable distances. Traces connected to all inductors are made wide enough to minimise series parasitic resistances and inductances, and thus avoid inductor Q degradation. Ideally, all interconnections should be as short as possible, to minimise the impact of parasitics. However, this is not always possible, especially due to the large geometrical structure of inductors when compared with components, such as transistors and

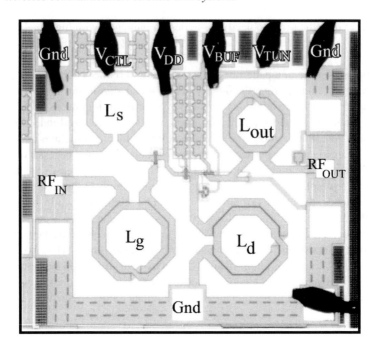

Figure 7.6   Micrograph of the 5.8 GHz CMOS LNA

resistors. When long interconnects become unavoidable in the layout, an in-house
interconnect modelling routine [26] is used. The main purpose of this routine is to
predict the additional parasitics introduced (e.g. inductances and resistances) and
account for them in simulation, increasing accuracy during the design phase. Finally,
line widths are set according to RF design guidelines, keeping DC traces thin and AC
connections wide and as short as possible.

7.2.8   *Measurement results – stand-alone LNAs*

7.2.8.1   Gain and frequency controllable sub-1 V, 5.8 GHz CMOS LNA

The 5.8 GHz LNA was implemented in a standard 0.18 μm CMOS process. It features
gain and frequency tuning capabilities. The experimental results were measured on
wafer using GGB Industries Inc. picoprobes and a 20 GHz Agilent 8720ES vector
network analyser. Standard short-open-load-through (SOLT) calibration procedures
were performed before taking measurements. The measured forward transmission
(S_{21}) is shown in Figure 7.7. For a power consumption of 16 mW from a 1 V supply,
a power gain of 13.2 dB is achieved at 5.8 GHz, with a noise figure of 2.5 dB. The
input and output reflection coefficients are less than −5 dB and −10 dB, respectively, as shown in Figure 7.8. The measured gain tuning of the LNA is shown in
Figure 7.9. A gain control of over 12 dB is achieved, without affecting the optimum
noise performance. The LNA exhibits a power gain greater than 7 dB at an extremely

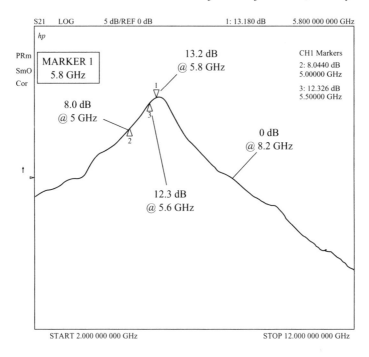

S21   LOG   5 dB/REF 0 dB   1: 13.180 dB   5.800 000 000 GHz

hp

PRm   ┌─────────────┐   13.2 dB   CH1 Markers
       │ MARKER 1   │   @ 5.8 GHz   2: 8.0440 dB
SmO   │ 5.8 GHz   │   5.00000 GHz
Cor   └─────────────┘
                                                                    3: 12.326 dB
       8.0 dB   5.50000 GHz
       @ 5 GHz

                                                    0 dB
                                                    @ 8.2 GHz

       12.3 dB
       @ 5.6 GHz

START 2.000 000 000 GHz   STOP 12.000 000 000 GHz

*Figure 7.7   Measured power gain of the single-ended 5.8 GHz CMOS LNA with a
1 V supply*

low voltage supply of 0.7 V (Figure 7.10), with a power consumption of 9.3 mW and
a noise figure of 2.68 dB.

A plot of the frequency tunability of the LNA is presented in Figure 7.11. A continuous frequency tuning of 360 MHz, from 5.6 GHz to 5.96 GHz, is achieved using
a simple varactor. This corresponds to a total tuning range of about 6 per cent, which
is smaller than the expected range of 10 per cent that was predicted from simulation.
An accumulation-type varactor was used. This structure has been reported to provide
a wider tuning range (i.e. 30 per cent) and a better quality factor, in comparison to
other PN or P$^+$N junction-based varactors. A stand-alone varactor structure, which
was fabricated on the same run as the LNA, was characterised. The measured capacitance tuning characteristics and quality factor of the varactor are shown in Figure 7.12.
The measured capacitance variation is about 19 per cent, which is smaller than the
expected 30 per cent. Furthermore, the measured quality factor of the varactor was
found to be much lower than predicted, and is in the order of $Q = 10$. These differences are mainly attributed to the inaccurate initial varactor modelling, especially
of the additive parasitic capacitances resulting after layout. The reduced tuning of
the varactor limits the frequency tuning of the LNA. The system was designed to
cover the complete 5–6 GHz frequency band, while the measured frequency tuning
is limited to 5.6–5.96 GHz. The unexpected low quality factor of the varactor also
has a direct impact on the resonant tank of the LNA. The estimated quality factor of

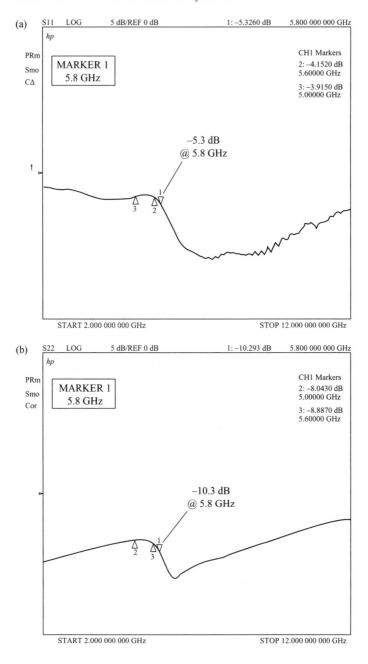

Figure 7.8   Measured (a) input and (b) output reflection coefficients of the 5.8 GHz CMOS LNA

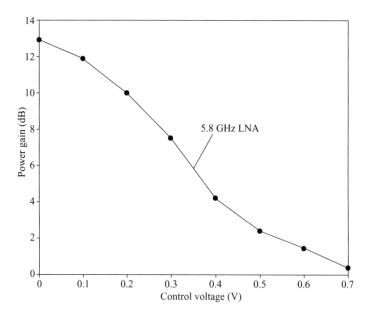

Figure 7.9   Gain tuning of the 5.8 GHz CMOS LNA

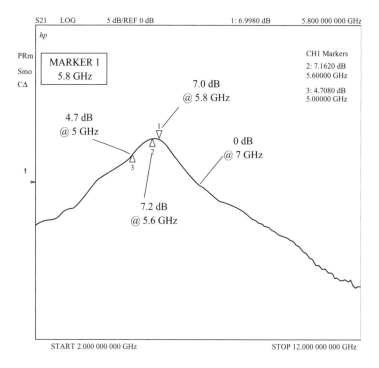

Figure 7.10   Measured power gain of the 5.8 GHz CMOS LNA with a power supply of 0.7 V

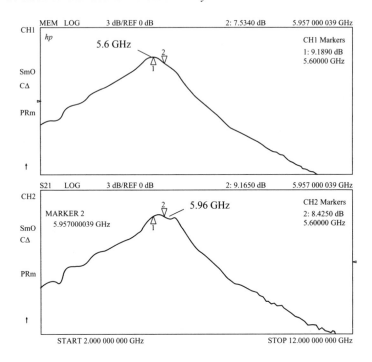

Figure 7.11   Frequency tuning of the 5.8 GHz CMOS LNA

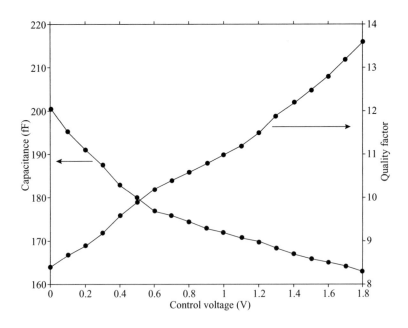

Figure 7.12   Capacitance tuning characteristics and quality factor of the varactors

the resonant tank (Q_{tank}) of the LNA is about $Q = 5.8$. This was found to be lower than the Q factors of similar LNA implementations which did not use varactors for frequency tuning.

7.2.8.2   Gain controllable sub-1 V, 8–9 GHz CMOS LNAs

Following the promising performance obtained from the 5.8 GHz LNA, two additional prototypes were fabricated, based on the same circuit topology, except for the absence of the tuning varactors. The microphotograph of the 9 GHz version is shown in Figure 7.13. The layout of the 8 GHz one is identical, except for the use of different inductor sizes.

The forward transmission (S_{21}) plots of both designs are shown in Figure 7.14. For a power consumption of around 20 mW from a 1 V supply, both prototypes achieved a power gain of 12–13.5 dB, with noise figures of 3.2–3.7 dB. Note that power gains of greater than 10 dB were achieved, over the frequency ranges of 6.7–8.6 GHz and 8.0–9.4 GHz, with the upper unity gain frequencies being at 10.8 GHz and 11.5 GHz, respectively. The input and output reflection coefficients of the two circuits are below −5 dB and −13 dB, respectively. Both LNAs exhibit power gains greater than 5 dB at an extremely low voltage supply of 0.7 V (Figure 7.15), for power consumptions of around 10 mW. A gain control of over 10 dB is achieved without

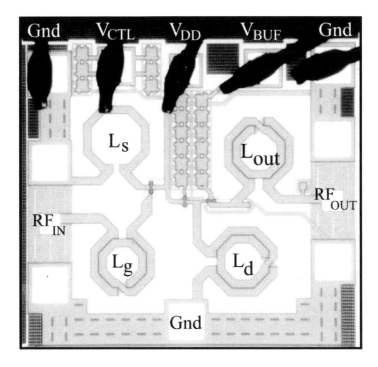

Figure 7.13   Micrograph of the 9 GHz CMOS LNA

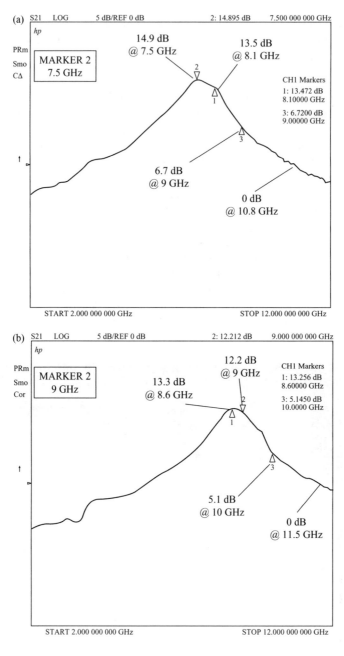

Figure 7.14   Measured power gain of the (a) 8 GHz and (b) 9 GHz CMOS LNAs

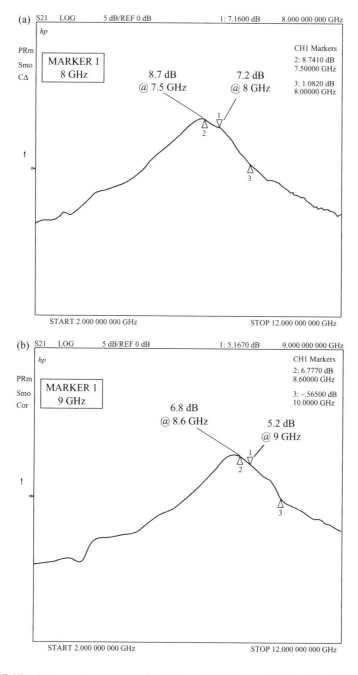

Figure 7.15 Measured power gain of the (a) 8 GHz and (b) 9 GHz CMOS LNAs with a power supply of 0.7 V

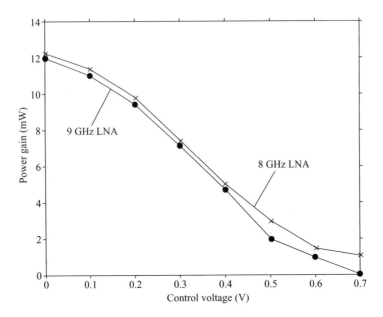

Figure 7.16   Gain tuning characteristics of the 8 and 9 GHz CMOS LNAs

any increase in circuit complexity. The gain tuning characteristics of both designs are shown in Figure 7.16.

It is interesting to note that, despite the higher operating frequency of the 9 GHz LNA, it has a relatively narrower bandwidth compared to the 8 GHz LNA. This is mainly due to the use of combined three metal layers (metal 4-5-6) for the inductors in the 9 GHz prototype, as opposed to the top two metal layers (metal 5-6) used in the 8 GHz circuit. This supports the fact that inductors with higher Q can be realised in CMOS by stacking more metal layers to emulate thicker conductors. The estimated quality factor of the resonant tanks (Q_{tank}) of the 8 GHz and 9 GHz CMOS LNA are 6.2 and 6.6, respectively. Recall that the Q_{tank} of the 5.8 GHz prototype was only 5.8 due to the use of varactors. The measured performances of all LNAs, under different biasing conditions, are summarised in Table 7.1.

7.3   A CMOS VCO architecture suitable for sub-1 volt GHz-range applications

Another key and critical building block in both wireless and optical communications transceivers is the voltage controlled oscillator (VCO). The continuous increase in the operating frequencies of integrated circuits, driven by the need for wider bandwidths and higher data rates, and the quest for system-on-chip solutions, resulted in a remarkable growth of interest in fully-integrated LC-based CMOS VCOs in recent years (e.g. [7, 10–12, 27–32]). Oscillating frequencies as high as 12.5 GHz have been

Table 7.1   LNA performance summary in this work

Design Technology	8 GHz LNA CMOS 0.18 μm		9 GHz LNA CMOS 0.18 μm		5.8 GHz LNA CMOS 0.18 μm	
V_{dd}	1 V	0.7 V	1 V	0.7 V	1 V	0.7 V
S_{21}	13.5 dB	7.1 dB	12.2 dB	5.2 dB	13.2 dB	7.0 dB
S_{11}/S_{22}	−5.8/−13.9 dB	−10.9/−17 dB	−5.4/−11.9 dB	−9/−12.9 dB	−5.3/−10.3 dB	−7.1/−12.3 dB
P_{dd}	22.4 mW	10.7 mW	19.6 mW	9 mW	16 mW	9.3 mW
NF	3.2 dB	4.1 dB	3.7 dB	4.7 dB	2.5 dB	2.68 dB
P_{in-1} dB	−13.2 dBm	−8.6 dBm	−8.9 dBm	−4.3 dBm	−14 dBm	−9 dBm
Gain tuning	11.4 dB	7.1 dB	11.2 dB	5.2 dB	12.6 dB	7.0 dB
Frequency tuning	–	–	–	–	360 MHz 5.6–5.96 GHz	

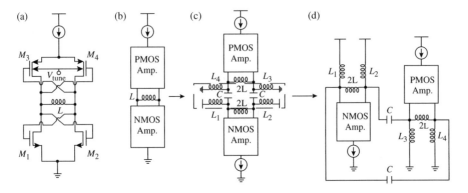

*Figure 7.17   (a)   Complementary   differential   LC   VCO   structure   [12];
(b)–(d) progression towards a low-voltage topology*

achieved using standard digital CMOS processes, e.g. [7, 12, 32]. The structure
of those VCOs employed stacked PMOS and NMOS transistors sharing the same
DC current (Figure 7.17a), and therefore requiring relatively high supply voltages
(2.5–3.5 V). Driven by the reduction of the power consumption of digital circuits and
the scaling of modern technologies, the supply voltages of integrated circuits continue to decrease towards sub-1 V. New circuit architectures are needed, especially
for analogue signal processors, to cope with this trend [33].

The VCO topology proposed in this section considerably reduces the supply
voltage requirement, and consequently the power consumption. This is done by
altering the structure of the conventional 'complementary differential LC circuit'
shown in Figure 7.17a [12, 28]. In addition to maintaining the features of the original topology (discussed in Section 7.3.1), the proposed architecture provides an
alternative to overcome the limited tuning range of back-gate tuning (Section 7.3.2).
Detailed circuitry and design guidelines for the proposed topology are presented in
Section 7.3.3.

Two VCO prototypes were implemented in a standard 0.18 μm CMOS process.
They operate using 0.85 and 1 V power supplies, which is approximately one third
the supply voltage needed by the original topology (labelled k and l in Figure 7.18).
This is achieved while satisfying other requirements such as low phase noise, low
power consumption (Figure 7.18), and a reasonable tuning range. Measured results
are reported and discussed in detail in Section 7.3.4.

7.3.1   The complementary differential LC structure

The complementary differential back-gate tuned VCO in Figure 7.17a has been shown
to allow very high frequencies of oscillation (9.8–12.5 GHz) [5, 32]. It uses NMOS and
PMOS cross-coupled amplifiers, along with a differential inductor L. The resonant
tank is formed by the inductor and the parasitic capacitances of both amplifiers.

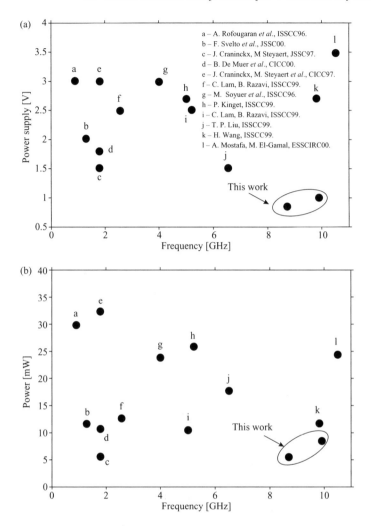

*Figure 7.18   Comparison of supply voltage and power consumption versus fre-
quency of state-of-the-art VCOs*

Frequency tuning can be performed by controlling the PMOS transistors' back-gate
voltages. This configuration has several desirable features:

(1)   The differential excitation of integrated inductors yields a higher effective
quality factor [34]. This results in a decrease in the VCOs phase noise.
(2)   The PMOS devices are used for frequency tuning by controlling their back-gate
voltages, thus eliminating the need for varactors which, as we have already
discussed earlier, tend to degrade the tank quality factor at high frequencies.
(3)   Dispensing with the varactors allows for higher frequency of oscillation.

(4)   The tank resonator is formed in a loop configuration, i.e. none of its elements is referenced to ground (refer to Section 7.3.3.1 and Figure 7.20). This makes the frequency of oscillation less sensitive to transistors' and inductors' parasitics to the substrate.

The main drawbacks of the topology in Figure 7.17a are the relatively high voltage supply required, and the limited tuning range provided by back-gate tuning.

7.3.2   Proposed low-voltage VCO circuit architecture

The progression of the circuit structure towards a low-voltage topology is shown in Figure 7.17. Capacitors are inserted between the PMOS and NMOS sections to decouple their DC bias, without affecting the AC interaction between the two tanks. Inductors L_1–L_2 are then added to ensure a DC path from the power supply to the NMOS tank, while presenting a high impedance to the AC signals. Similarly, inductors L_3–L_4 secure a DC path for the PMOS tank to ground.

It is clear that the resulting topology significantly reduces the voltage supply required, while maintaining the characteristics of the original circuit. A second advantage is gained by decoupling the DC biasing of the two tanks. Tuning can now be done either using the back-gate voltage of the PMOS tank, or via its bias current. Combining both tuning mechanisms results in a wider tuning range (see Figure 7.23 on page 177).

7.3.3   VCO detailed circuitry and design guidelines

Figure 7.19 shows the complete transistor-level VCO circuit. In the following subsections, we briefly highlight the main design issues related to the proposed structure.

7.3.3.1   The LC tank resonator

Figure 7.20a shows the AC equivalent circuit of the LC tanks of the VCO in Figure 7.19. It is composed of two differential inductors $2L$, in parallel with a total capacitance composed of the drain–gate capacitances C_{dg} and gate–source capacitances C_{gs} of M_1–M_4. All the capacitances related to M_1–M_2 form the fixed component of the tank capacitance, whereas those of M_3–M_4 form the variable part. The design requirements of the DC-blocking capacitors C, and the AC-blocking inductors L_1–L_4, are discussed in subsequent sections. In the high-frequency equivalent circuit shown in Figure 7.20b, the coupling capacitors are treated as short circuits, and L_1–L_4 as open circuits. The oscillator frequency is given by $f_0 = 1/(2\pi\sqrt{L_{tank}C_{tank}})$, where

$$C_{tank} = C_{dgM1} + C_{dgM2} + \frac{(C_{gsM1} + C_{gsM2})}{C_{gsM1}C_{gsM2}} + C_{dgM3} + C_{dgM4}$$

$$+ \frac{(C_{gsM3} + C_{gsM4})}{C_{gsM3}C_{gsM4}}, \tag{7.7}$$

and $L_{tank} = 2L\|2L = L$.

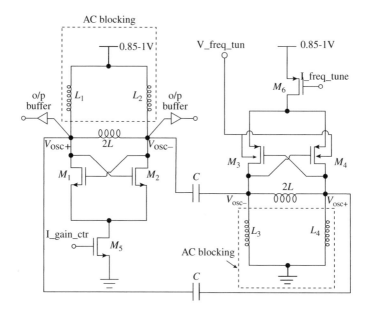

Figure 7.19   Circuit of the proposed VCO using capacitively coupled NMOS-PMOS LC tanks. The output buffers are on-chip PMOS transistors with 50 Ω resistor loads. V-freq-tun: Back-gate control voltage for frequency tuning. I-freq-tune: Bias-current control voltage for frequency tuning

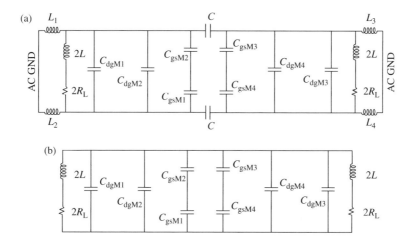

Figure 7.20   (a) Tank components for the circuit in Figure 7.19, and (b) high frequency equivalent circuit

7.3.3.2   The coupling capacitors

The sizes of capacitors C need to be chosen to ensure they present a low-impedance path to the RF signal between the PMOS and NMOS tanks, at the frequency of oscillation. For this, the following condition needs to be satisfied

$$2C \gg C_{tank}. \tag{7.8}$$

We used two 2.5 pF high-quality metal-insulator-metal (MIM) capacitors, which represent approximately 7–8 Ω of resistance at f_0.

7.3.3.3   The AC-blocking inductors

The role of L_1–L_4 is to prevent the tank's energy from leaking to the voltage supplies. They should act as AC-blocking impedances. The impedance of each one of those inductors at f_0, Z_{Lblock}, needs to be significantly larger than the tank impedance Z_{tank}. It can be shown that this condition is ensured when the following relations are satisfied

$$\frac{Z_{Lblock}}{2} \gg Z_{Ltank} \| Z_{Ctank}, \tag{7.9}$$

$$L_{block} \gg \frac{2L_{tank}}{1 + (4\pi^2 f_0^2 L_{tank} C_{tank})}. \tag{7.10}$$

In the prototypes presented here, each one of the blocking inductors is made up of a combination of an integrated spiral inductor in series with the bonding inductance of the package, with a total of approximately 3.5 nH.

7.3.4   *VCO experimental results*

A micrograph of the 8.7 GHz VCO chip is shown in Figure 7.21. Note that symmetry is conserved throughout the entire layout. All control nodes are bonded for packaging, while ground-signal-ground pads were used for on-chip probing of the RF output. The tank's coupling capacitors are built using the three top metal layers only, in order to avoid signal leakage to the substrate. Unit parallel capacitor cells were used to ensure good matching.

The measured single-ended output spectra for the 8.7 and 10 GHz circuits are shown in Figure 7.22, and were used to directly estimate the phase noise. The latter are −86 dBc/Hz and −82 dBc/Hz at a 100 kHz offset for the 8.7 and 10 GHz VCOs, respectively. Note that the average power of the signal is underestimated due to the narrow span and limited resolution bandwidth of the spectrum analyser. In estimating the phase noise, the signal power was measured using a wider span [10].

Two tuning mechanisms can be used to vary the VCO's frequencies. Figure 7.23a shows the frequency tuning using the back-gate voltages of the PMOS tank at a constant bias current of 2.36 mA. Figure 7.23b shows frequency tuning using the bias current of the PMOS tank at a constant back-gate voltage of 0.85 V. The two diamond points in the top figure show the minimum and maximum achievable frequencies,

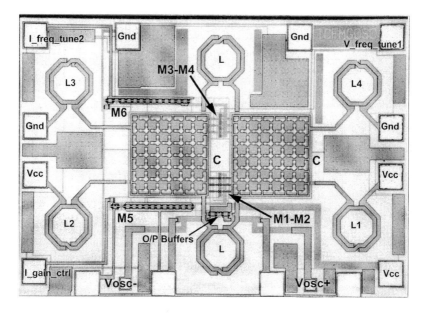

Figure 7.21   Micrograph of the 8.7 GHz VCO

when both tuning approaches are combined. A maximum tuning range of 400 MHz is measured for the 8.7 GHz prototype, and 450 MHz for the 10 GHz.

One added advantage of the proposed topology is that the output amplitude is not very sensitive to the PMOS tank bias current, since the transconductance is mainly provided by the NMOS tank, while relatively lower current is used in the PMOS tank. Figure 7.24 shows the measured output power of the 8.7 GHz VCO as both the back-gate voltage and bias current of the PMOS transistors are varied. The measurements indicate that as the PMOS tank bias current varies from 0 mA to 4 mA, the output power changes only by about 3.5 dBm, which can be a worthy trade-off for the increased tuning range. Table 7.2 summarises the characteristics and performance of the two VCO prototypes.

In order to fairly compare the overall performance of the VCOs presented here to that of state-of-the-art CMOS implementations operating at different frequencies and power levels, we make use of the following figure of merit (FOM) [35]

$$\text{FOM} = L_{\text{measured}}(\Delta f) - 20\log(f_{\text{osc}}/\Delta f) + 10\log(P_{\text{dissip}}/1\text{ mW}), \qquad (7.11)$$

where f_{osc} is the frequency of oscillation, Δf is the phase noise measured at an offset from f_{osc}, and P_{dissip} is the DC power dissipation in the VCO. Equation (7.11) assumes a phase noise slope of 20 dB/decade, as suggested by Leeson's oscillator noise model [36]. A summary of the comparison result is shown in Figure 7.25. While achieving a very high frequency of oscillation at the lowest supply voltages, the FOMs

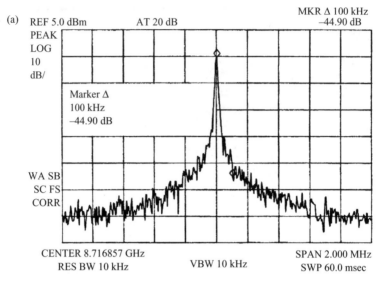

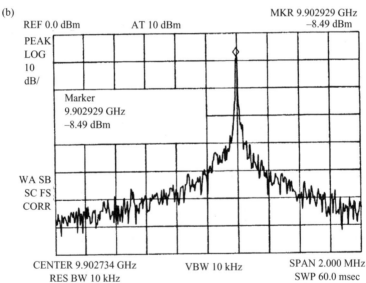

Figure 7.22   Measured single-ended outputs of the (a) 8.7 GHz and (b) 10 GHz VCOs

of the two VCOs compare favourably to other implementations. Note that the signal power and phase noise of the VCOs labelled d and e in Figure 7.25 were measured differentially [28, 31], as opposed to the single-ended measurements of most of the other implementations. Note also that the VCO labelled k [12] requires more than

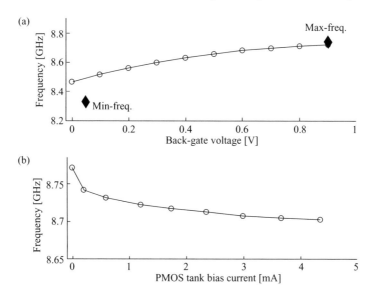

Figure 7.23   Frequency tuning versus (a) back-gate voltage, and (b) PMOS tank bias current. The two diamond points on the top figure show the minimum and maximum achievable frequencies, when both tuning schemes are combined

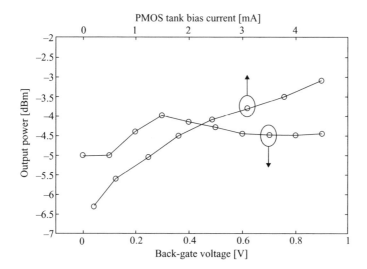

Figure 7.24   Output power variation of the 8.7 GHz VCO as a function of both the back-gate and bias current of the PMOS tank

Table 7.2   Summary of the process characteristics, and performances of the two VCOs

	Prototype 1 [8.7 GHz]	Prototype 2 [10 GHz]
Technology	0.18 µm, 6 metal standard CMOS	0.18 µm, 6 metal standard CMOS
Substrate resistivity	10 Ω/cm	10 Ω/cm
Thickness of top metal	0.99 µm	0.99 µm
Differential inductor	1 nH	0.85 nH
Estimated quality factor of inductor at f_0	4	5
Size of M_1-M_2	100 µm	100 µm
Size of M_3-M_4	100 µm	50 µm
Size of PMOS buffer	50 µm	50 µm
Chip area	1.5 mm × 1.1 mm	1.5 mm × 1.1 mm
Supply voltage	0.85 V	1 V
Supply current	7.1 mA	9 mA
Power consumption	6.0 mW	9 mW
Phase noise @ 100 kHz offset	−86 dBc/Hz	−82 dBc/Hz
Phase noise @ 600 kHz offset	−100 dBc/Hz	−98 dBc/Hz
Phase noise @ 1 MHz offset	−103 dBc/Hz	−101 dBc/Hz
Tuning range	400 MHz	450 MHz
Tuning sensitivity	470.6 MHz/V	450 MHz/V

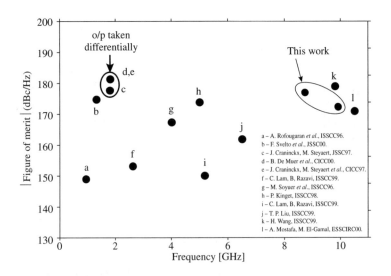

Figure 7.25   Comparison of the figure of merits of the VCOs presented in this work to that of state-of-the-art

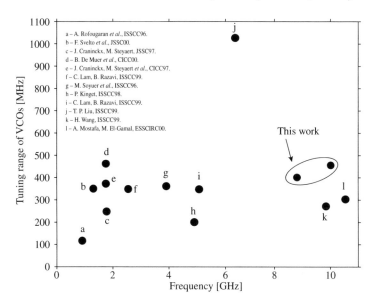

Figure 7.26   Comparison of proposed VCOs tuning ranges to state-of-the-art

2.5 times the supply voltage of the VCOs reported here. Finally, a comparison of the tuning ranges to those of state-of-the-art implementations is given in Figure 7.26.

7.4   A very low-voltage (0.8 V) 5 GHz CMOS receiver frontend

Traditionally, RF receivers employed heterodyne architectures, using discrete filters for image rejection and channel selection. However, as the demand for higher integration increases, the direct conversion architecture has gained popularity, due to its simplicity and ease of integration. In this scheme, the RF signal is directly converted to baseband. Consequently, no image reject filter is required after the first amplification stage, and therefore the LNA does not need to drive off-chip high-quality filters.

Figure 7.27 shows the complete architecture of the receiver presented in this chapter, with the differential low-noise amplifier (LNA) connected to a set of I/Q mixers which are driven by a voltage-controlled oscillator (VCO) with quadrature outputs. The outputs of the I/Q mixers can subsequently be connected to another down-conversion stage to translate the IF signal to baseband, and thus implementing a Weaver image reject architecture (e.g. [37]). Another alternative is to directly process the IF signal, after being converted to single-ended by a transformer, using a high-speed ADC [38] followed by a digital signal processor, as suggested in Figure 7.27. The outputs of the mixers are capacitively coupled to the following stages in order to mitigate the effect of DC offset, resulting from LO leakage and signal feedthrough.

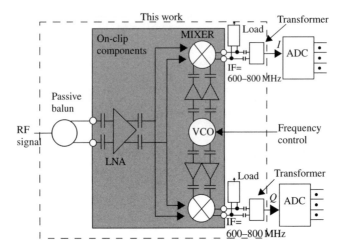

Figure 7.27   Receiver architecture

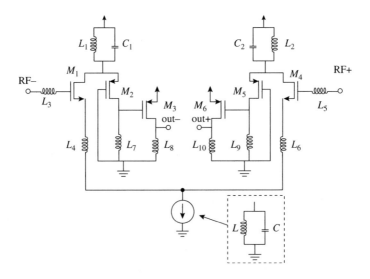

Figure 7.28   The differential LNA used in the receiver

7.4.1   The differential LNA

Figure 7.28 shows a simplified schematic of the differential LNA used in the receiver, including all on-chip inductors. It is the differential version of the LNA discussed earlier in this chapter. In order to avoid added complexity and risk, the tuning capabilities discussed for the single-ended version were not explored in the receiver. Transistors M_1 and M_4 act as common-source amplifiers, and transistors M_2 and M_5 are used as common-gate current buffers with load inductors L_7 and L_9. Note that an added advantage of this structure is that, due to the higher voltage headroom, the transistors can be biased deeper into saturation, leading to an improved linearity. Also, this topology has shown improved reverse isolation.

The LC tanks, formed by inductors L_1 and L_2 and capacitors C_1 and C_2, behave as DC current sources. They provide the necessary DC bias current to the LNA, while presenting a high impedance to the RF signals. Also, a positive by-product of using an LC tank as such is to nullify the effect of the parasitic capacitances of the transistors, resulting in an improvement of the noise figure [14]. At resonance, the low quality factor of an on-chip inductor can cause the impedance of the LC tank to be small compared to the impedance seen at the sources of the PMOS transistors M_2 and M_5. As a result, the RF signal could be lost to the LC tank [15], thus the importance of ensuring the quality of the inductors used. In order to further maximise the voltage headroom, the biasing current source was also implemented as a resonating tank.

In order to achieve minimum noise, the widths of transistors M_1 and M_4 need to be set to their optimum values, W_{opt}. According to [39]

$$W_{\text{opt}} = \frac{1}{3\omega L R_s C_{\text{ox}}},\tag{7.12}$$

where ω is the operating frequency of the circuit in radian/sec, L is the minimum feature size of the fabrication process (0.18 μm in this work), C_{ox} is the gate capacitance per unit area, and R_S is the input resistance (50 Ω in this case). Inductors L_3 and L_4 are used for matching as explained earlier. Buffers formed using transistors M_3 and M_6 provide extra gain and increased isolation between the LNA and the mixer.

7.4.2   The mixer

The mixer (Figure 7.29), which is based on a double-balanced Gilbert active structure, down-converts the differential RF signal from 5.2 GHz to a 600–800 MHz intermediate frequency (IF). It consists of a pair of transconductors (M_5–M_6) and four transistors (M_1–M_4) which act as switches.

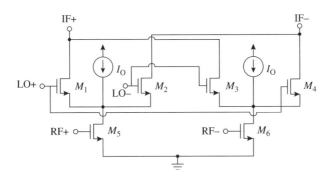

Figure 7.29   Schematic of the I/Q mixer

Although passive mixers generate less noise than active mixers, they require larger LO signals and provide no conversion gain, resulting in a severe noise figure deterioration. On the other hand, while active mixers can provide signal gain, they suffer from the noise induced by the transconductor and switching transistors. To mitigate the noise from the latter, which appears directly in the output signal path, extra current sources I_0 are used as shown in Figure 7.29. This decreases the DC biasing currents of M_1–M_4, thus reducing their noise, while maintaining the high bias of transistors M_5–M_6 necessary to ensure a high RF gain [40].

To improve the mixer linearity, the differential input transconductor was built as a grounded-source differential pair, rather than a differential pair with constant current source biasing. A current source biasing would introduce third-order intermodulation products and lower the overall linearity of the mixer [41]. This also helps reducing the voltage headroom requirement.

Since the mixer is to operate from a 0.8 V supply, transistors with relatively large widths are used, in order to lower the threshold voltage required to bias the transistors in saturation. In addition, the gates of transistors M_1–M_4 are biased near threshold in order to minimise their switching time, thus reducing the output noise. The RF signals from the LNA are AC coupled to the inputs of the mixers, in order to allow the latter to set different DC biasing points. As for the loads at the IF outputs, they can either be resistive or reactive (e.g. an LC tank).

7.4.3   The quadrature voltage-controlled oscillator

The core structure of the VCO (Figure 7.30) consists of two pairs of cross-coupled transistors M_1–M_2 and M_3–M_4 with on-chip load inductors L_1–L_4. The two transistor pairs are coupled through transistors M_5–M_8 in order to provide quadrature outputs for both the I and Q paths [30]. Inductors L_1–L_4 resonate with the parasitic capacitors of the transistors to set the oscillation frequency, while the transistors provide the negative resistances necessary to overcome the inductor losses.

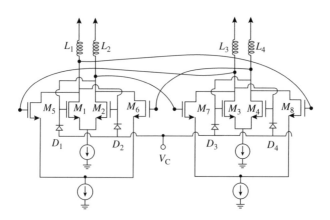

Figure 7.30   Schematic of the quadrature VCO

The output phase noise at an offset Δf from f_0 can be approximated by the following relationship

$$L\{\Delta f\} = kT(1 + A)Z_0 \frac{1}{Q_{\text{tank}}} \left(\frac{f_0}{\Delta f}\right)^2 \frac{1}{V_{\text{rms}}^2}, \tag{7.13}$$

where kT is the product of the Boltzmann constant with the absolute temperature, A is the noise factor safety margin necessary to ensure oscillation start-up, V_{rms} is the root mean square voltage at the oscillation nodes, and $Z_0 = \sqrt{(L/C_{\text{tank}})}$ and Q_{tank} are the tank characteristic impedance and quality factor, respectively. In order to ensure oscillation, the widths of the transistors in our design were slightly oversized to increase the safety margin A.

Varactors D_1–D_4, which are implemented as NMOS transistors over Nwells, are controlled with an external DC voltage source, V_c, in order to provide tuning capabilities. As a result, the tuning range of the VCO is around 5 per cent, from 4.4 GHz to 4.6 GHz. The size of the varactors is 100 µm. Larger varactors could be used to provide a wider tuning range, while adjusting the transistor sizes accordingly. Also, switched arrays of capacitors could be used for coarse tuning. Common-source buffers with inductive loads are used to increase isolation to the mixer, and to provide very desirable amplification for the VCO signals driving the mixers.

7.4.4 Measured characteristics of the receiver

The receiver was fabricated in a standard 0.18 µm CMOS technology from TSMC. Figure 7.31 shows a photomicrograph of the chip which occupies an area of 3.2 mm × 1.7 mm.

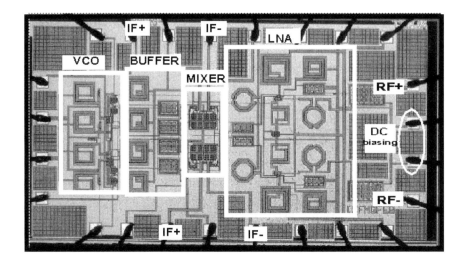

Figure 7.31   Receiver chip photomicrograph

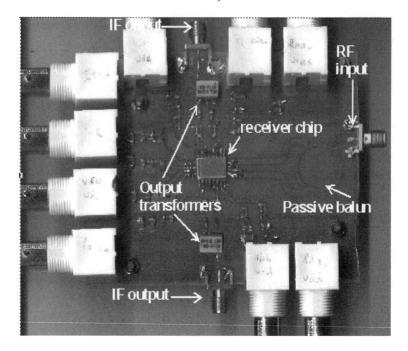

Figure 7.32   Printed circuit board test setup for the receiver

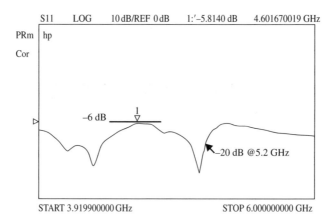

Figure 7.33   Input reflection S_{11} of the receiver

A custom test fixture was built to characterise the fully packaged circuit (Figure 7.32). A passive ring hybrid balun structure on PCB is used at the input to convert the single-ended RF signal to a differential signal. A sliding capacitor, which slides along the two parallel transmission lines that connect the balun to the receiver chip, was used to match the input impedance of the receiver to 50 Ω. Figure 7.33 shows a plot of the resulting input reflection S_{11} of the receiver.

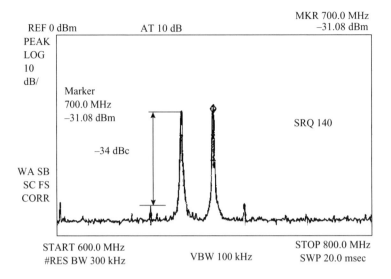

Figure 7.34   *Receiver output spectrum for a two-tone test response, without calibrating out the test set-up losses*

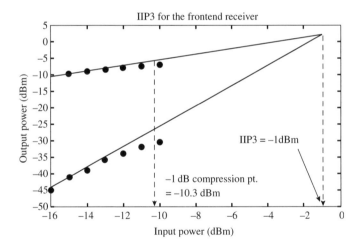

Figure 7.35   *Receiver measured third-order intercept plot, after calibrating out the test set-up losses*

Resistive loads of 50 Ω are used at the mixer outputs. A surface mount transformer is then used to convert the output differential IF signal to single-ended for measurement via SMA connectors.

Figure 7.34 shows the measured output spectrum, without calibrating out the test set-up losses, for a two-tone test input at 5.2 GHz and 5.22 GHz. The fundamental signals appear at 680 MHz and 700 MHz, while the third-order intermodulation outputs,

Table 7.3   Measured receiver performance

Overall conversion gain (50 Ω system)	6 dB
VCO tuning range	4.4–4.6 GHz
VCO phase noise	−100 dBc/Hz @ 500 kHz
Overall noise figure	7 dB
Image rejection	22 dB
Input S_{11} @ 5.2 GHz	−20 dB
IIP3	−1 dBm
Input 1-dB compresssion	−10.3 dBm
Power dissipation @ 0.8 V	56 mW
Die size (0.18 μm CMOS)	5.44 mm^2

which are located at 660 MHz and 720 MHz, are −34 dBc below the fundamentals. The same set-up was used to obtain the overall conversion gain, the 1 dB compression point, and the IIP3 of the receiver. Figure 7.35 shows the third-order intercept plot, after calibrating out the losses of the test set-up (e.g. baluns, transformers, etc.). Table 7.3 summarises the measured receiver performance.

7.5   Conclusions

This chapter has demonstrated the potential of using standard digital state-of-the-art submicron CMOS technologies to realise critical radio frequency integrated circuit building blocks, operating in the gigahertz range. In order to cope with the trend of technology down-scaling, and the continuous decrease in the supply voltages of digital circuitry, the work focused on circuit architectures and topologies suitable for operation from 1 V supplies and lower, while delivering performances comparable, and in some respects, better than similar circuits requiring higher supply voltages.

In particular, a folded cascode LNA structure was shown to be very versatile. It allows gain tuning without any additional circuit complexity or performance degradation. Similarly, a very low voltage VCO structure that maintains all the features of the conventional complementary differential LC VCO circuit, while operating from a much lower supply, and providing a wider tuning range, was proposed. Finally, a complete 5 GHz receiver frontend that was optimised for operation from a supply voltage as low as 0.8 V, and that features high image rejection, was presented.

All the architectures and circuits proposed were demonstrated through measurements from different IC prototypes. It should be noted that the CMOS technology we used did not offer any special RF oriented devices, such as thick metal layers, etc. In practice, that would be the case, suggesting that the performances reported in this chapter are expected to be further improved. Finally, this chapter discussed practical integrated inductor design guidelines and layout techniques, which are both crucial for the success of any RF circuit implementation.

7.6 References

1 MOMOSE, H. S., MORIFUJI, F., YOSHITOMI, T., OHGURO, T., SAITO, M., MORIMOTO, T., KATSUMA, Y. and IWAI, H.: 'High frequency AC characteristics of 1.5 nm gate oxide MOSFETs'. Proceedings of IEEE International Electron Device Meeting (IEDM) San Francisco, CA, USA, Dec. 1996, pp. 105–08

2 TSANG, T. K. K. and EL-GAMAL, M. N.: 'Gain controllable very low voltage (1 V) 8–9 GHz integrated CMOS LNA's'. Radio Frequency Integrated Circuits (RFIC) Symposium Digest of Technical Papers, Seattle, WA, USA. IEEE, Jun. 2002, pp. 205–08

3 TSANG, T. K. K. and EL-GAMAL, M. N.: 'Gain and frequency controllable sub-1 V 5.8 GHz CMOS LNA'. Proceedings IEEE International Symposium on *Circuits and Systems (ISCAS)*, Arizona, USA, May 2002, pp. 795–98

4 LEROUX, P. and STEYAERT, M.: 'High-performance 5.2 GHz LNA with on-chip inductor to provide ESD protection', *Electronic Letters*, Mar. 2001, **37** (7), pp. 467–68

5 MOSTAFA, A. H. and EL-GAMAL, M. N.: 'A CMOS VCO architecture suitable for sub-1 volt high-frequency (8.7–10 GHz) RF applications'. IEEE International Symposium on Low Power Electronics and Design (ISLPED), Huntington Beach, CA, USA, Aug. 2001, pp. 247–50

6 MOSTAFA, A. H. and EL-GAMAL, M. N.: 'A sub-1V 4 GHz CMOS VCO and a 12.5 GHz oscillator for low-voltage and super-high frequency applications', *IEEE Transactions on Circuits and Systems II (TCAS II)*, Oct. 2001, **48** (10), pp. 919–26

7 MOSTAFA, A. H. and EL-GAMAL, M. N.: 'A fully integrated sub-1 V 4 GHz CMOS VCO, and a 10.5 GHz oscillator'. 26th European Solid-State Circuits Conference (ESSCIRC), Stockholm, Sweden, IEEE, Sept. 2000, pp. 312–15

8 DE RANTER, C. and STEYAERT, M.: 'A 0.25 µm CMOS 17 GHz VCO'. IEEE ISSCC Digest Technical Papers, San Francisco, CA, USA, Feb. 2001, pp. 370–71

9 KINGET, P.: 'A fully integrated 2.7 V 0.35 µm CMOS VCO for 5 GHz wireless applications', IEEE ISSCC Digest Technical Papers, San Francisco, CA, USA, Feb. 1998, pp. 226–27

10 LAM, C. and RAZAVI, B.: 'A 2.6 GHz/5.2 GHz CMOS VCO', IEEE ISSCC Digest Technical Papers, San Francisco, CA, USA, Feb. 1999, pp. 402–03

11 LIU, T.: 'A 6.5 GHz monolithic CMOS VCO'. IEEE ISSCC Digest Technical Papers, San Francisco, CA, USA, Feb. 1999, pp. 404–05

12 WANG, H.: 'A 9.8 GHz back-gate tuned VCO in 0.35 µm CMOS'. IEEE ISSCC Digest Technical Papers, San Francisco, CA, USA, Feb. 1999, pp. 406–07

13 LEE, K. and EL-GAMAL, M. N.: 'A very low-voltage (0.8 V) CMOS receiver frontend for 5 GHz RF applications'. Proceedings IEEE International Symposium on *Circuits and Systems (ISCAS)*, Arizona, USA, May 2002, pp. 125–28

14 SAMAVATI, H., RATEGH, H. R. and LEE, T. H.: 'A 5-GHz CMOS wireless LAN receiver front-end', *IEEE Journal of Solid-State Circuits*, May 2000, **35** (5), pp. 765–72

15 ABOU-ALLAM, E., NISBET, E. and MALIEPAARD, M. C.: 'Low-voltage 1.9 GHz front-end receiver in 0.5-μm CMOS technology', *IEEE Journal of Solid-State Circuits*, Oct. 2001, **36** (10), pp. 1434–43

16 TSANG, T. K. K. and EL-GAMAL, M. N.: 'A fully integrated 1 V 5.8 GHz bipolar LNA'. Proceedings of IEEE International Symposium on *Circuits and Systems (ISCAS)*, Sydney, Australia, May 2001, pp. 843–45

17 HUNG, C. M. and KENNETH, K. O.: 'A packaged 1.1 GHz CMOS VCO with phase noise of −126 dBc/Hz at 600-kHz offset', *IEEE Journal of Solid-State Circuits*, Jan. 2000, **35** (1), pp. 100–103

18 PENNISI, S., SCACCIANOCE, S. and PALMISANO, G.: 'A new design approach for variable-gain low noise amplifiers'. Radio Frequency Integrated Circuits (RFIC) Symposium Digest of Technical Papers, Boston, MA, USA, IEEE, Jun. 2000, pp. 139–42

19 SACCHI, E., BIETTI, I., GATTA, F., SVELTO, F. and CASTELLO, R.: 'A 2 dB NF, fully differential, variable gain, 900 MHz CMOS LNA'. Symposium on VLSI Circuits Digest of Technical Papers, Honolulu, HI, USA, IEEE, Jun. 2000, pp. 94–97

20 RAFLA, R. A. and EL-GAMAL, M. N.: '2.4–5.8 GHz CMOS LNA's using integrated inductors'. IEEE Midwest Symposium on Circuits and Systems, Lansing, MI, USA, Aug. 2000, **1**, pp. 302–04

21 SORRAPANTH, T., YUE, C., SHAEFFER, D., LEE, T. and WONG, S.: 'Analysis and optimization of accumulation-mode varactor for RF ICs'. Symposium on VLSI Circuits Digest of Technical Papers, Honolulu, HI, USA, Jun. 1998, IEEE, pp. 32–33

22 CASTELLO, R., ERRATICO, P., MANZINI, S. and SVELTO, F.: 'A +/− 30% tuning range varactor compatible with future scaled technologies'. Symposium on VLSI Circuits Digest of Technical Papers, Honolulu, HI, USA, IEEE, Jun. 1998, pp. 34–35

23 KOUTSOYANOPOULOS, Y., PAPANANOS, Y., BANTAS, S. and ALE-MANNI, C.: 'Performance limits of planar and multi-layer integrated inductors'. Proceedings of IEEE International Symposium on *Circuits and Systems (ISCAS)*, Geneva, Switzerland, May 2000, pp. 160–63

24 NIKNEJAD, A. and MEYER, R.: 'Analysis, design, and optimization of spiral inductors and transformers for Si RF ICs', *IEEE Journal of Solid-State Circuits*, Oct. 1998, **33** (10), pp. 1470–81

25 MOSTAFA, A. H., LEE, K. and EL-GAMAL, M. N.: 'Characterization of CMOSP18 inductors', Technical Report, Canadian Microelectronics Corporation, Ottawa, Canada, Jan. 2001

26 ZHANG, H.: 'High frequency parasitic modeler', VLSI Design Project, McGill University, Sept. 2001

27 SVELTO, F., DEANTONI, S. and CASTELLO, R.: 'A 1.3 GHz low-phase noise fully tuneable CMOS LC VCO', *IEEE Journal of Solid-State Circuits*, Mar. 2000, **35** (3), pp. 356–61

28 CRANINCKX, J., STEYAERT, M. and MIYAKAWA, H.: 'A fully-integrated spiral-LC CMOS VCO set with prescaler for GSM and DCS-1800 systems'.

Proceedings of Custom Integrated Circuits Conference (CICC), Santa Clara, CA, USA, IEEE, May 1997, pp. 403–06

29 CRANINCKX, J. and STEYAERT, M.: 'A 1.8-GHz low-phase-noise CMOS VCO using optimized hollow spiral inductors', *IEEE Journal of Solid-State Circuits*, May 1997, **32** (5), pp. 736–44

30 ROFOUGARAN, A., RAEL, J., ROFOUGARAN, M. and ABIDI, A.: 'A 900 MHz CMOS LC-Oscillator with Quadrature Outputs'. IEEE ISSCC Digest Technical Papers, San Francisco, CA, USA, Feb. 1996, pp. 392–93

31 DE MUER, B., ITOH, N., BORREMANS, M. and STEYAERT, M.: 'A 1.8-GHz highly tuneable low-phase-noise CMOS VCO'. Proceedings of Custom Integrated Circuits Conference (CICC), Orlando, FL, USA, IEEE, Feb. 2000, pp. 585–88

32 MOSTAFA, A. H. and EL-GAMAL, M. N.: 'A 12.5 GHz back-gate tuned CMOS voltage controlled oscillator'. IEEE International Conference on Electronics, Circuits and Systems (ICECS), Jounieh, Lebanon, Dec. 2000

33 SANSEN, W., STEYAERT, M., PELUSO, V. and PEETERS, E.: 'Toward Sub 1 V Analog Integrated Circuits in Submicron Standard CMOS Technologies'. IEEE ISSCC Digest Technical Papers, San Francisco, CA, USA, Feb. 1998, pp. 186–87

34 DANESH, M., LONG, J. R., HADAWAY, R. A. and HARAME, D. L.: 'A Q-factor enhancement technique for MMIC inductors', Radio Frequency Integrated Circuits (RFIC) Symposium Digest of Technical Papers, Baltimore, MD, USA, IEEE, Jun. 1998, pp. 217–20

35 BALTUS, P. G. M., WAGEMANS, A. G., DEKKER, R., HOOGSTRAATE, A., MAAS, H., TOMBEUR, A. and VAN SINDEREN, J.: 'A 3.5-mW, 2.5-GHz diversity receiver and a 1.2-mW, 3.6-GHz VCO in silicon on anything', *IEEE Journal of Solid-State Circuits*, Dec. 1998, **33** (12), pp. 2074–79

36 LEESON, D. B.: 'A simple model of feedback oscillator noise spectrum'. Proceedings of the IEEE, Feb. 1966, pp. 329–30

37 RUDELL, J. C., OU, J.-J., CHO, T. B., CHIEN, G., BRIANTI, F., WELDON, J. A., GRAY, P. R.: 'A 1.9 GHz wide-band IF double conversion CMOS receiver for cordless telephone applications', *IEEE Journal of Solid-State Circuits*, Dec. 1997, **32** (12), pp. 2071–88

38 COPELAND, M. A., VOINIGESCU, S. P., MARCHESAN, D., POPESCU, P., MALIEPAARD, M. C.: '5-GHz SiGe HBT monolithic radio transceiver with tuneable filtering'. IEEE Transactions on Microwave Theory and Techniques, Feb. 2000, **48**, pp. 170–81

39 SHAEFFER, D. K. and LEE, T. H.: 'A 1.5-V, 1.5-GHz CMOS low noise amplifier', *IEEE Journal of Solid-State Circuits*, May 1997, **32** (5), pp. 745–59

40 RAZAVI, B.: 'RF Microelectronics' (Prentice Hall, 1998)

41 WU, S. and RAZAVI, B.: 'A 900-MHz/1.8-GHz CMOS receiver for dual-band applications', *IEEE Journal of Solid-State Circuits*, Dec. 1998, **33** (12), pp. 2178–85

Chapter 8

Design of integrated CMOS power amplifiers for wireless transceivers

Mona M. Hella, Seoung-Jae Yoo and Mohammed Ismail

8.1 Introduction

The continuing reduction in production costs and improvements in personal communication systems has led to a tremendous growth in the wireless communications market. For cheap wireless terminals, it is attractive to integrate the RF frontend with the backend signal processing to reduce assembly cost. Until now, power amplifiers for wireless applications were produced almost exclusively in GaAs technologies, with a few exceptions in LDMOS, Si BJT and SiGe HBT. While CMOS provides high functionality and complexity at low costs, for an RF power amplifier, the problem of using CMOS technology is more severe than other blocks in the transceiver due to the limited voltage-handling capability. The linearity and power efficiency are lower than other technologies. Therefore, implementation of RF power amplifiers in CMOS has been one of the most challenging tasks for wireless transmitters. The design of power amplifiers in CMOS technology is mainly affected by the following factors.

1   Low breakdown voltage of deep-submicron technologies. This limits the maximum gate–drain voltage since the output voltage at the transistor's drain normally reaches 2 times the supply for classes B and F, and around 3 times the supply for class E operation. Thus, transistors have to operate at a lower supply voltage, delivering lower power. Additionally CMOS technology has lower current drive; i.e. the gain provided by the single stage is very low. Either multiple stages would be used or new design techniques, which would reduce the number of stages by decreasing the input drive requirements of the large transistors in the power amplifier (PA), are employed [1].

2   In contrast to semi-insulating substrates, a highly doped substrate is common in CMOS technology. This results in substrate interaction in a highly integrated

CMOS IC. The leakage from an integrated power amplifier might affect the stability of, for example, the VCO in a transceiver chain.

3   Conventional transistor models for CMOS devices have been found to be moderately accurate for RF ICs, and need to be improved for analogue operation at radio frequencies. Large-signal CMOS RF models and substrate modelling are critical to the successful design and operation of integrated CMOS radio-frequency power amplifiers, owing to the large currents and voltage changes that the output transistors experience [2]. As a result, traditional PA design relies heavily on data measured from single transistors.

4   Since the inherent output device impedance in the power amplifier case is very low, impedance matching becomes very difficult, requiring higher impedance transformation ratios. Additionally, output matching elements require lower loss, and good thermal properties since there are usually significant RF currents flowing in these elements. If CMOS technology is used, losses in the substrate will decrease the quality factor of passive elements in the matching network. Usually the output-matching network is implemented off-chip as the antenna itself is off-chip.

5   The power amplifier delivers large output current in order to achieve required power at the load. This current can be high enough that electro-migration and parasitics in the circuit may cause performance degradation [2].

The goal of this chapter is to provide the main concepts and challenges of RF power amplifier designs in CMOS technology. In particular, the circuit techniques and analysis of power amplifiers for portable applications are discussed.

8.2   Power amplifiers (PAs): concepts and challenges

Power amplifiers (PAs) are part of the transmitter frontend, and are used to amplify the signal being transmitted so that it can be received and decoded within a fixed geographical area. The design of PAs, especially for linear, low-voltage operation, is still a difficult task. In practice, PA design has involved a substantial amount of trial and error, with discrete and hybrid implementations being traditionally used. The main performance parameters for power amplifiers are the level of output power it can achieve, depending on the targeted application, linearity and efficiency. There are two basic definitions for the efficiency of a PA. These are drain efficiency and power-added efficiency (PAE). The drain efficiency is the ratio between the RF output power to the DC consumed power, and the PAE is the ratio between the difference of the RF output power and the RF input power to the DC consumed power. The PAE is a more practical measure as it accounts for the power gain of the amplifier. As the power gain decreases, more stages will be required. Since each stage consumes a certain amount of power, the overall power consumption increases, decreasing the overall efficiency. While power efficiency is a performance issue, linearity is imposed by the utilised modulation technique, or by the level of output power back-off during operation.

8.2.1 Conjugate match and load-line match

The concept of conjugate match is widely known as setting the value of the load impedance equal to the real part of the generator's impedance such that maximum output power is delivered to the load. However, the delivered power is limited by the maximum rating of the transistor acting as a current generator, together with the available supply voltage.

By referring to Figure 8.1, it is evident that the device in this case would show limiting action at a current considerably lower than its physical maximum of I_{max}. This means that the transistor is not being used to its full capacity. To utilise the maximum current and voltage swing of the transistor, a load resistance of lower value, commonly referred to as the load-line match, R_{opt} is used. In its simplest form, it is defined as $R_{opt} = V_{max}/I_{max}$, assuming the generator's resistance is much higher than the optimum load resistance. Thus the load-line match represents a real compromise that is necessary to extract the maximum power from the RF transistor, and at the same time to keep the RF voltage swing within the specified limits of the transistor and the available DC supply.

Figure 8.2 illustrates the effect of the difference of gain match versus power (load-line) match on the output of a linear amplifier. The solid line shows the response of an amplifier that has been conjugate matched at much lower drive levels. The two points A and B refer to the maximum linear power and the 1 dB compression power.

In a typical situation, the conjugate match yields a 1 dB compression power about 2 dB lower than that which would be obtained by the correct power tuning, shown by the dotted line in Figure 8.2. This means that the device would deliver 2 dB lower power than the device specification of the manufacturer. Since in power amplifier design, it is always required to extract the maximum possible from a transistor, the power-matched condition has to be taken more seriously, despite the fact that the gain at lower signal levels may be 1 dB or less than the conjugate-matched condition. Across a wide range of devices and technologies, the actual difference in output

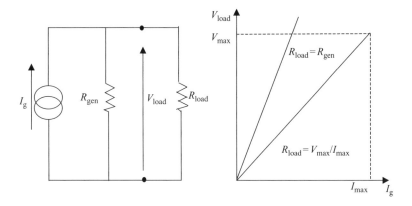

Figure 8.1   Conjugate match and load-line match

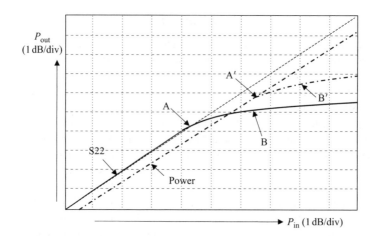

Figure 8.2 *Compression characteristics for conjugate match (S22) (solid curve) and power match (dotted curve). 1 dB gain compression points (B′, B) and maximum power points (A′, A) show similar improvements under power-matched conditions*

power, gained by the power-matched condition, may vary over a range of 0.5 dB to 3 dB [3].

However, a load-line (power) match rather than a conjugate (gain) match might cause reflections and a voltage standing wave ratio (VSWR) in a system to which it is connected. The reflected power is entirely a function of the degree of match between the antenna and the 50 ohm system. The PA does present a mismatched reverse termination, which could be a problem in some situations. An isolator or a balanced amplifier [4] is a simple and effective way of dealing with the problem.

8.2.2 Effect of the transistor knee (pinch-off) voltage

Traditional power amplifier design starts by determining the optimum load using the load-line approach as shown in Figure 8.3. The knee voltage (pinch-off voltage) divides the saturation and the linear region of the transistor and can be defined as V_{ds} (voltage between drain and source) at the 95 per cent of I_{max} point.

The optimum load resistance is

$$R_{opt} = \frac{(V_{max} - V_{knee})}{I_{max}}. \tag{8.1}$$

While this is an effective approach for most power transistors, it is not suitable for submicron CMOS transistors. This is mainly due to the fact that V_{knee} is only about 10–15 per cent of the supply voltage for typical power transistors, while it can be as high as 50 per cent of the supply for deep-submicron technologies. Therefore, precluding the CMOS transistor from operating in the linear region does not result in optimum output power. In fact, a large portion of the RF cycle can be in the linear

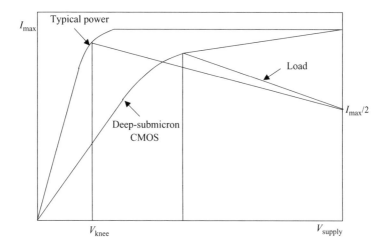

Figure 8.3   Effect of the knee voltage on the determination of the optimum load

region. Therefore, both saturation and the linear region must be considered when determining the optimum load. This can be done using a general MOSFET equation valid in all regions of operation [5] or relying on harmonic balance simulations of circuits, with accurate transistor models, as will be discussed later.

8.2.3   Classification of power amplifiers

Power amplifiers have been traditionally categorised under many classes: A, B, C, AB, D, E, F, etc. [6]. Power amplifier classes can be categorised either as bias point dependent, such as classes A, B, AB and C, or as dependent on the passive elements in the output matching network that shape the drain voltage and current, with the transistor, in this case, operating as a switch. The choice of operating class of the power amplifier is largely determined by the wireless standard utilised. For example, many wireless standards that are located in the 900 MHz band or close to it, such as GSM, NADC (835 MHz), and other applications that use the ISM band do not require a high degree of linearity. A class E power amplifier can be employed for GSM and applications that use the ISM band, with some form of added linearisation for NADC application. In the next subsection, details of each operating class are discussed.

8.2.3.1   Class A, B, AB and C power amplifiers

The primary distinction between these power amplifier classes is the fraction of the RF cycle for which the transistor conducts. For class A PAs, the transistor is conducting for the entire RF cycle, whereas for class B PAs it is ON for half the RF cycle, and for less than half the RF cycle for class C. Class A, AB and B amplifiers may be used as linear PAs, while class C is more non-linear in nature [7].

Figure 8.4 illustrates the schematic and the associated current waveforms for the above-mentioned classes of operation. While the third-order intercept point (IP3),

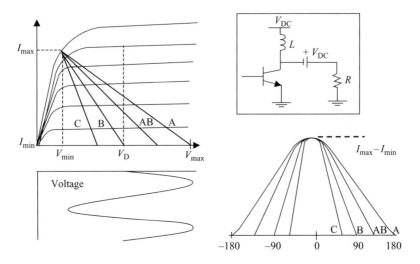

Figure 8.4   Traditional illustration of the schematic and associated current wave-forms of classes A, B, AB and C

adjacent channel power ratio (ACPR), 1 dB compression point, and harmonics are various measures of linearity of PAs, drain efficiency and power added efficiency (PAE) of the PA are used to indicate the current drawn from the supply. The PAE is defined as

$$\text{PAE} = \frac{P_{\text{rf,out}} - P_{\text{rf,in}}}{P_{\text{dc}}} \tag{8.2}$$

where $P_{\text{rf,out}}$ is the RF output power, $P_{\text{rf,in}}$ is the RF input power, and P_{dc} is the total DC power drawn from the supply.

The efficiency and output power for a power amplifier operating in class A, AB, B, or C, are given by [3]

$$\eta = \frac{V_{\text{dd}} - V_{\text{dsat}}}{V_{\text{dd}}} \frac{\theta - \sin \theta}{4(\sin(\theta/2) - (\theta/2)\cos(\theta/2))} \tag{8.3}$$

$$P_{\text{out}} = \frac{1}{2}(V_{\text{dd}} - V_{\text{dsat}})\frac{I_{\text{m}}}{2\pi}(\theta - \sin \theta) \tag{8.4}$$

where V_{dd} is the supply voltage, θ is the conduction angle of the drain current, V_{dsat} is the pinch-off voltage (knee voltage), and I_{m} is the maximum drain current in the input transistor. Equations (8.3) and (8.4) are plotted in Figure 8.5a. From this figure, it is evident that an increase in efficiency, obtained by reducing the conduction angle, is achieved at the expense of the reduced output power from the power amplifier. In deep-submicron technologies, the low output power of a reduced conduction angle is a major drawback. In order to achieve the required output power, load resistance has to be lowered to impractical values comparable to values of the parasitic resistances.

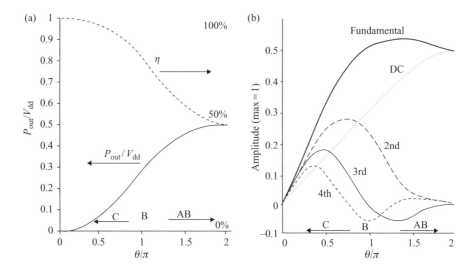

Figure 8.5   *(a) RF power and efficiency as a function of the conduction angle; (b) Fourier analysis of the drain current*

As the conduction angle of the drain current decreases, the harmonic content of the current signal increases. The magnitude of the nth harmonic of the output drain current is given by [3]

$$I_n = \frac{1}{\pi} \int_{-\theta/2}^{\theta/2} \frac{I_m}{1 - \cos(\theta/2)} \left(\cos\alpha - \cos\left(\frac{\theta}{2}\right) \right) \cos(n\alpha)\, d\alpha. \qquad (8.5)$$

By examining Figure 8.5b, it is clear that the DC component decreases monotonically as the conduction angle is reduced. In class B, the fundamental component is the same as in class A while the DC component is reduced by $\pi/2$. For conduction angles below π, corresponding to class C operation, the DC component continues to drop, but the fundamental component of the current signal also starts to drop below its class A level. This results in high efficiency, and lower power utilisation factor (PUF). The odd harmonics can be seen to pass through zero at the class B point. For class AB mode, the third harmonic is not negligible. Still, class AB represents a compromise between linearity, PUF and efficiency.

8.2.3.2   Class D

The class D PA is a switching mode PA, and has recently been implemented in CMOS technology. For an efficient amplifier, it would consist of a controlled switch, in which the on-resistance was zero, the off-resistance was infinite and the transition time was zero. Then the output signal would consist of the power supply switched at the rate of the input signal, with no losses in the switching device [8]. A class D PA uses the switching configuration of the MOS transistor to achieve high efficiency.

In the class D PA, current from the supply is steered between the device, when the switch is closed, and the load, when the switch is open [9]. According to the duty cycle of switching, some fraction of the input voltage is amplified to create an output voltage. If the switching is done at the output carrier frequency, the narrowband nature of the transmitted signal allows the use of RF filters to pass only the fundamental frequency component. Because of the ability to filter out unwanted components of the output signal, this type of amplification can be done with only one device, in which case the power from the supply is either sunk in the device or the load [9]. The use of a series LC circuit tunes to the output frequency and the current in the device will be a sinusoid for the period in which it conducts current [9]. In case two devices are used, each will carry a half-sinusoidal current waveform. The ideal efficiency of a class D PA is 100 per cent if the on-resistance and the output voltage are zero at the time of the close of the switch. However, this maximum efficiency cannot be obtained due to the non-zero on-resistance and finite transition time of the switch [9].

The choice of class in PAs depends on several factors. Switching-mode power amplifiers such as class D, E and F have generally higher efficiency than linear power amplifiers because an ideal switch does not have an overlapped period of non-zero switch voltage and current. Practically, however, the transition between the ON and OFF state of a switch takes a finite time, during which a substantial amount of power can be dissipated as shown in Figure 8.6. This kind of switching is called hard switching, and is one of the main reasons for efficiency reduction in switching-mode power amplifiers such as class D and F. On the other hand, the load network in class E power amplifiers is designed such that the switch voltage returns to zero with zero slope right before the switch turns on, ensuring no overlap of non-zero switch voltage and current. Figure 8.7 shows the voltage and current waveforms of soft switching. This soft switching of the class E power amplifier minimises the power loss in the switch and thus highly efficient amplification is possible for constant envelope modulated signals [10].

As the supply voltage decreases, the value of the optimum load required to achieve a specific value for output power also decreases. This decrease in the load

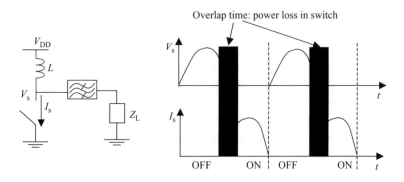

Figure 8.6   *Waveforms of a switching-mode power amplifier with hard switching*

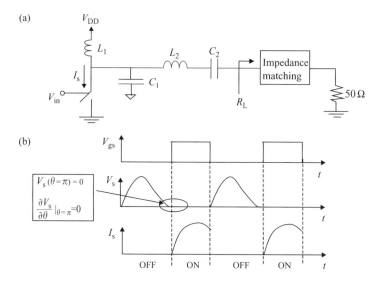

Figure 8.7 (a) Typical schematic of a class E power amplifier; (b) its voltage and current waveforms showing the soft switching characteristics

resistance will increase the matching network transformation ratio from 50 ohm, caus-ing more losses in the matching network. This effect is less pronounced in class E than classes B [11], C and F [10, 12], making it more suitable for low-voltage operation.

8.2.3.3 Class E

Figure 8.8 shows a conceptual picture of a class E power amplifier [13, 14]. The input signal V_{in} toggles the switch periodically with approximately 50 per cent duty cycle. When the switch is ON, a linearly increasing current is built up through the inductor. At the moment the switch is turned off, this current is steered into the capac-itor, causing the voltage across the switch V_s to rise. The tuned network is designed such that in steady state V_s returns to zero with a zero slope, immediately before the switch is turned on. The bandpass filter then selectively passes the fundamental component of V_s to the load, creating a sinusoidal output that is synchronised in phase and frequency with the input. In practical applications, V_{in} may be phase or frequency modulated, in which case the information embedded in the modulation is also phased to the output with power amplification [1]. By comparing V_s, and I_s in Figure 8.8, it can be observed that the switch voltage and current are never simultaneously non-zero. Since the instantaneous power dissipation of the switch is the product of these two quantities, the switch is ideally lossless, and all the power from the DC supply is delivered to the radio frequency output. In addition, the capacitor is designed to be fully discharged before the switch is turned ON.

In high-speed operation, the switch transition time can become a significant fraction of the signal period. During these transitions, the switch voltage and current

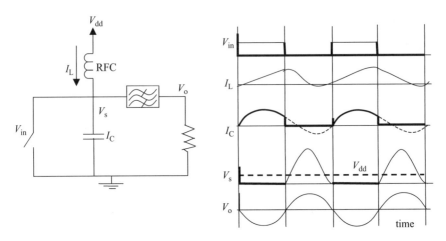

Figure 8.8   A simplified class E power amplifier, and its steady state operation

may be simultaneously non-zero, causing potential power loss in typical switching amplifiers. For proper class E operation, this loss is alleviated at the turn-on transistors by a zero switch current resulting from a simultaneously zero V_s and dV_s/dt. On the other hand, turn-off transition loss is reduced by delaying the switch voltage rise until the switch is turned off. The properties have made class E PAs attractive for high-efficiency operations.

One of the features of class E amplifiers is the large peak voltage that the switch sustains in the off state, approximately $3.56V_{dd} - 2.56V_{min}$, where V_{min} is the minimum voltage across the transistor. Operating at class E requires either a high transistor breakdown voltage, or operating at V_{dd} less than the specified value for a given technology. Figure 8.9 shows the basic schematic of a class E stage, with the associated values of circuit components.

The formulas describing the dependence of various elements on output power (P_{out}), supply voltage (V_{dd}), loaded quality factor (Q_L), and operating frequency ($\omega = 2\pi f$) are derived in [15], based on the following assumptions:

1   The inductance of the DC choke is very high.
2   The quality factor of the series inductor (L) is high.
3   The losses in the switch are negligible.

By utilising the class E conditions, $v_D(\pi) = 0$ and $i_D(\pi) = 0$, and a 100 per cent power efficiency assumption, the drain voltage waveform of the amplifier becomes

$$v_D = \frac{V_{DD}}{\pi}\left(\theta + \frac{\pi}{2}\cos\theta + \sin\theta - \frac{\pi}{2}\right) \qquad (8.6)$$

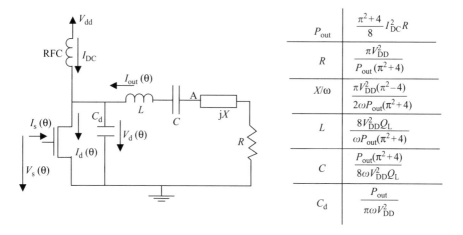

P_{out}	$\dfrac{\pi^2+4}{8}I_{DC}^2 R$
R	$\dfrac{\pi V_{DD}^2}{P_{out}(\pi^2+4)}$
X/ω	$\dfrac{\pi V_{DD}^2(\pi^2-4)}{2\omega P_{out}(\pi^2+4)}$
L	$\dfrac{8V_{DD}^2 Q_L}{\omega P_{out}(\pi^2+4)}$
C	$\dfrac{P_{out}(\pi^2+4)}{8\omega V_{DD}^2 Q_L}$
C_d	$\dfrac{P_{out}}{\pi\omega V_{DD}^2}$

Figure 8.9   *Single-ended class E resonant power amplifier*

where $\theta = \omega t$. The drain voltage waveform is Fourier transformed in order to solve the fundamental frequency phase angle ϕ_1 and amplitude α_1 of the signal at node A:

$$\tan\phi_1 = -\frac{\int_0^1 v_D \cos\theta\, d\theta}{\int_0^1 v_D \sin\theta\, d\theta} \tag{8.7}$$

$$\alpha_1 = \frac{1}{\pi}\sqrt{\left(\int_0^\pi v_D \sin\theta\, d\theta\right)^2 + \left(\int_0^\pi v_D \cos\theta\, d\theta\right)^2}. \tag{8.8}$$

To achieve the correct phase at the load, an excess reactance X is added in series with the load resistance. The values for R and X are calculated using the solved α_1 and ϕ_1:

$$R = \frac{v_{out}^2}{2P_{out}} = \frac{\alpha_1^2}{2P_{out}}\left(1+\left(\frac{\pi\tan\phi_1+2}{\pi-2\tan\phi_1}\right)^2\right)^{-1} \tag{8.9}$$

$$X = R\tan\left(\phi_1 - \tan^{-1}\left(\frac{-2}{\pi}\right)\right). \tag{8.10}$$

The maximum drain voltage occurs when $\theta = 2\tan^{-1}(2/\pi)$:

$$v_{D,max} = 2\pi V_{DD}\tan^{-1}\left(\frac{2}{\pi}\right) = 3.562 V_{DD}. \tag{8.11}$$

8.2.3.4   Class F

The basic idea behind class D and F is to shape the output signal at the drain of the transistor such that it has more of a square shape than a sinusoidal shape. The load network provides a high termination impedance at the second or third harmonics. Therefore the voltage waveform across the switch exhibits sharper edges than a sinusoid, thereby lowering the power loss in the transistor.

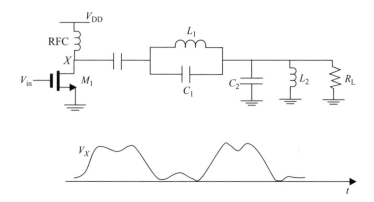

Figure 8.10   Schematic diagram and output waveform of a typical class F stage

Figure 8.10 shows an example of a class F topology. A tank consisting of L_1 and C_1 resonates at either $2f_{in}$ or $3f_{in}$, where f_{in} is the input signal frequency, thus boosting the second or third harmonics at point X. The voltage across the switch approaches a rectangular waveform as the third harmonic becomes stronger. If the drain current of M_1 is assumed to be a half-sinusoid (i.e. half-wave rectified sinusoid), then it contains no third harmonic. The product of a rectangular drain voltage and half-wave rectified current represents the power losses in the transistor. Since the power losses are minimum due to the shaping of the two signals, the efficiency can be relatively high. The theoretical efficiency of a class F power amplifier can reach 85 per cent.

To summarise the discussion of PA classes, what determines the class of operation of the power amplifier is its conduction angle, input signal overdrive, and the output load network. Figure 8.11 shows how the PA relates to the conduction angle and the input signal overdrive. It illustrates that a given PA can be in any of the classical operating modes depending on the above two factors. For a small RF input signal V_{in}, the amplifier can operate in class A, AB, B or C depending on the conduction angle (bias voltage relative to the transistor's threshold voltage). The PA efficiency can be improved by reducing its conduction angle and moving the design into class C operation, but at the expense of lower output power. An alternative approach to increasing efficiency without sacrificing output power is to increase the input overdrive such that the transistor acts as a switch. These are called saturated class A and C, class D, class E, or class R, depending on the conduction angle, and the shape of the load network.

8.2.4   Power amplifier linearisation

Linearisation techniques are mostly utilised in basestations due to their complexity. For mobile phones, increasing the talk time and lowering the weight of the terminal rely on having an efficient amplifier that does not consume a lot of DC power. On the other hand, an efficient amplifier is normally non-linear, while a spectrally efficient

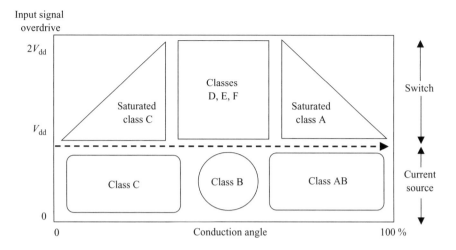

Figure 8.11   Classical definition of power amplifier classes

modulation technique produces non-constant envelope signals. If this non-constant envelope signal is applied to a non-linear amplifier, the signal suffers spectral growth, which leads to adjacent channel interference. One of the solutions would be to use an efficient non-linear PA and apply a suitable linearisation technique to restore linearity.

The conventional linearisation techniques are feedforward, feedback predistortion [3, 11], envelope elimination and restoration (EER) [16], linearisation using non-linear components (LINC) [17], bias adaptation, and Doherty amplifier [3]. The first three techniques are complex and need adjustments, or pre-measured data to achieve the required linearisation and are used in basestations. The simplicity of the last four techniques makes them amenable to integration depending on the degree of linearity required and the channel bandwidth. Even for modulation techniques that do not require linearisation, techniques like EER, LINC, Doherty's amplifier, and bias adaptation can be used for efficiency enhancement at lower output power levels.

8.2.4.1   Feedforward

A non-linear power amplifier generates an output voltage waveform that can be viewed as the sum of a linear replica of an input signal and an error signal. A feedforward topology computes this error and with proper scaling subtracts it from the output waveform. Shown in Figure 8.12 is a simple example where the output of the main PA, V_M, is scaled down by $1/A_V$, generating V_N. The input is subtracted from V_N, and the result is scaled by A_V and subtracted from V_M. We note that if $V_N = V_{in} + V_D/A_V$, yielding $V_P = V_D/A_V$, and $V_Q = V_D$, then $V_{out} = A_V V_{in}$. In practice, the two amplifiers in the circuit exhibit substantial phase shift at high frequencies, mandating the use of delay lines such that Δ_1 compensates for the phase shift of the PA, and Δ_2 for the phase shift of the error amplifier.

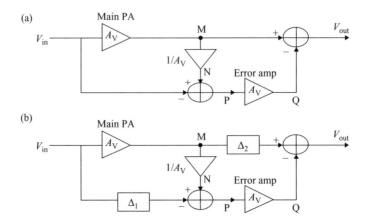

Figure 8.12   *(a) Simple feedforward topology, (b) addition of delay elements*

The advantage of feedforward topologies over feedback methods is inherent stability even with finite bandwidth and substantial phase shift in each building block. This is particularly important in RF and microwave circuits because inevitable poles and resonances at frequencies near the band of interest make it difficult to achieve stable feedback. Feedforward linearisation elements require passive devices such as microstrip lines, with the power loss of Δ_2 being critical. The output subtractor must be realised using a low loss component such as a high-frequency transformer [18].

8.2.4.2   Doherty amplifier

The Doherty amplifier is primarily an efficiency enhancer rather than a linearisation technique. It employs relatively linear amplifiers, which are known to have lower efficiency at lower power levels. It is used to preserve the peak efficiency at back-off points in modulation schemes that have high peak to power ratio. This means that for a given level of linearity, or spectral regrowth, a desired level of mean RF power can be achieved using the same device periphery but at substantially higher efficiency than in simple open-loop configuration.

The principle of the Doherty amplifier is to use one main power amplifier (PA) and one auxiliary PA. At maximum output power, both PAs contribute equally to the output. Upon decreasing the input drive level until typically half the maximum combined output power (-6 dB from P_{max}), the auxiliary PA approximately shuts down. The high efficiency of the Doherty amplifier is achieved by keeping the main amplifier at maximum device output voltage when the auxiliary amplifier is operating. The high device output voltage results in high power efficiency. The schematic of the Doherty amplifier and the corresponding output power waveforms are shown in Figure 8.13.

The Doherty amplifier uses what is called the active load pull technique, which means that the whole operation is equivalent to resistance or reactance of the RF load being modified by applying current from a second phase-coherent source, which is

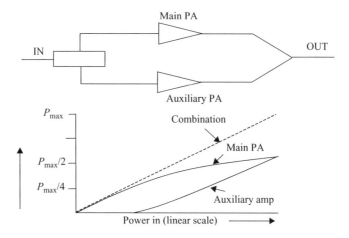

Figure 8.13   Basic Doherty amplifier configuration

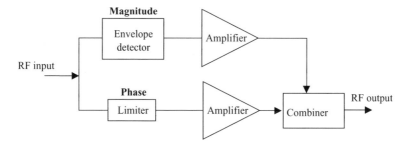

Figure 8.14   Conceptual diagram of envelope elimination and restoration technique

the auxiliary amplifier. By doing so, the impedance seen by different amplifiers is a function of other elements and the common load. The load-pulling effect together with a quarter-wave transformer causes the effective load resistance to decrease with increasing drive level. This impedance transformation is necessary to keep the main amplifier device voltage at its maximum in the high-power region.

The power efficiency of the main amplifier alone is ideally constant in the high-power region. The auxiliary amplifier has its highest power efficiency at maximum output power. Therefore the complete Doherty amplifier has a high efficiency in the whole power range, especially at medium output power compared to classic power amplifier designs.

8.2.4.3   Envelope elimination and restoration

Figure 8.14 shows the block diagram of the EER (envelope elimination and restoration) linearisation scheme as proposed by Khan [16]. As the name EER implies, the envelope of the RF input is first eliminated by a limiter to generate a constant amplitude phase signal. At the same time, the magnitude of input information is extracted by an envelope detector.

The magnitude and phase are amplified separately, and then recombined to restore the desired RF output. A way to combine the magnitude and phase components is to use an efficient switched-mode RF power amplifier. In a switched mode PA, output power is directly proportional to the square of the supply voltage. Thus, the envelope of the RF output of a switched-mode RF PA is directly proportional to its supply voltage. Envelope and phase components can therefore be recombined if the phase signal (RF) is applied to the gate of a transistor and the magnitude signal (low frequency) directly modulates the supply. The key advantage of this EER approach is that the PA always operates as an efficient switched-mode amplifier. Thus, high efficiency can be obtained without compromising linearity.

In practice, the process of amplifying the detected envelope signal up to the necessary voltage and current capacity to modulate the PA device consumes a significant amount of power. However, modern techniques centring in high-efficiency pulse width modulation developed for high fidelity audio amplification can be used for this application, maintaining a relatively high efficiency [19]. One problem would be the bandwidth of the modulator, which would appear to be limited to a few megahertz. This limits the use of this technique to certain modulation standards.

8.2.4.4   Linear amplification using non-linear components

This technique is also known as the out-phasing amplifier. It adopts the same concept as EER in the use of non-linear power amplifiers, but avoids non-constant envelope input signals.

According to Figure 8.15, an input signal $v_{in}(t) = a(t) \cos[\omega_c t + \phi(t)]$ can be expressed as a sum of two constant-amplitude phase-modulated signals, $v_1(t) = 0.5V_0 \sin[\omega_t t + \phi(t) + \theta(t)]$ and $v_2(t) = 0.5V_0 \sin[\omega_t t + \phi(t) - \theta(t)]$ where $\theta = \sin^{-1} a(t)/V_0$.

Thus, if $v_1(t)$ and $v_2(t)$, generated from $v_{in}(t)$, are amplified by means of non-linear stages, and subsequently added, then the output signal will contain the same

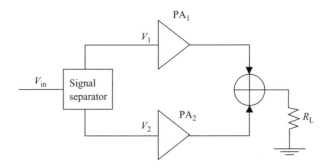

Figure 8.15   Linear amplification using non-linear stages

envelope and phase information as $v_{in}(t)$. Realisation of $v_1(t)$ and $v_2(t)$ from $v_{in}(t)$ requires substantial complexity, primarily because their phase must be modulated by $\theta(t)$, which itself is a non-linear function of $a(t)$. The separation of varying envelope signals into two constant-envelope signals is referred to as signal component separation (SCS). Recently reported work on integrated 200 MHz SCS has demonstrated a sideband suppression of 45 dBc using two open-loop, amplitude-compensated saturated wideband BJT amplifiers [20]. In practice, LINC transmitters must deal with two critical issues. First, the gain and phase mismatch between the two signal paths as shown in Figure 8.15 results in residual distortion. Second, the interaction between the non-linear amplifiers through the combiner network limits the overall linearity achieved in this open-loop configuration as the two non-linear amplifiers when connected together might cause the two phase modulated signals to corrupt each other's phase. Nevertheless, this kind of transmitter has aroused wide interest lately, and reported results have shown good linearity at 1 GHz [21].

8.2.5 Spectral regrowth

Abrupt changes in a digitally modulated waveform, for example, QPSK, result in envelope variation if a filter limits the bandwidth of the signal before being applied to the PA. If the power amplifier exhibits significant non-linearity, then the shape of the input signal to be transmitted is not preserved, and the spectrum is not limited to a desired bandwidth. This effect is called spectral regrowth, and can be quantified by the relative adjacent channel power. Figure 8.16 illustrates this effect in the case of a QPSK signal applied to a weakly non-linear power amplifier.

In order to limit spectral regrowth, linear power amplifiers are usually utilised. However, linear PAs are usually less efficient than non-linear ones since they consume

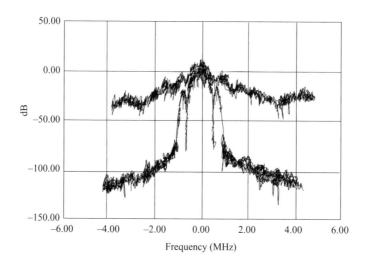

Figure 8.16   Spectral regrowth due to amplifier non-linearity

Table 8.1   Example of some digital wireless standards

Parameter	NADC	IS-95 CDMA	GSM	Bluetooth	IEEE802.11
RF Tx. Freq. (MHz)	824–849	1860–1910	890–915	2400–2497	2400–2497
Multiple access	TDMA/FDM	CDMA/FDM	TDMA/FDM	Frequency hopping	DSSS
Duplexing	FDD	FDD	FDD	TDD	TDD
Modulation	$\pi/4$-DQPSK	QPSK/OQPSK	GMSK	GFSK	DQPSK
Peak to average ratio	3.5 dB	10 dB	1.5 dB	0 dB	0 dB
Spectral regrowth	Medium	High	Low	Low	Medium

a considerable amount of power with respect to the rest of the portable phone. Non-linear PAs, on the other hand, exhibit efficiencies as high as 60 per cent. Thus, it is desirable to employ modulation schemes that do not experience spectral regrowth when processed by non-linear amplifiers.

Table 8.1 shows basic modulation techniques of some standards. Techniques employing $\pi/4$-DQPSK and QPSK/OQPSK require a highly linear power amplifier to limit the spectral growth caused by their abrupt phase changes. Although standards employing GMSK, FM and GFSK do not require high linearity, some standards like GSM have power control mechanisms that necessitate efficiency enhancement techniques at lower power levels. Another feature required of power amplifiers in digital wireless standards is control of output power. For example, in TDMA systems such as IS-54 and GSM, the PA is turned on and off periodically to save power. Also in IS-95, the output power must be variable in steps of 1 dB. In class 1 Bluetooth radio, the output power must be controlled from 4 dBm to 20 dBm in steps of 2, 4, 6 or 8 dB.

8.3   Power amplifier controllability and stability

Implementing power amplifiers in CMOS technology is considered a major step towards realisation of a complete transceiver on-chip. Modern transceivers require means for adjusting the transmitted power over a finite range to further reduce power consumption and improve channel capacity. A low-performance, short-range wireless standard such as Bluetooth requires a high level of integration and low-cost that can only be achieved using CMOS technology together with means of controlling the output power up to 20 dBm. In addition, power amplifiers are typically backed off relative to their peak power and power added efficiency points in order to meet the linearity requirement of the system. The degree of back-off varies depending on the modulation scheme employed: 0 dB for Gaussian-filtered minimum shift keying (GMSK for GSM and DECT), 7 dB for $\pi/4$-DQPSK (IS-54 and PHS), 10 dB for

QPSK (IS 95) and 12 dB for 16 QAM being typical figures. Thus adding efficient techniques for adjusting the output power level is considered a challenging issue to integrated power amplifiers. Along with controllability of RF PAs, the stability is an important design issue. RF PA oscillation problems can be broadly categorised into two kinds: bias oscillations and RF oscillations.

8.3.1 Power amplifier controllability

Linear power amplifiers can have their power adjusted by variation of biasing or by a dynamic variation of load seen by the output stage (Doherty amplifier [3]). Output power can also be controlled through a variation of the input signal amplitude, realised by having a variable gain amplifier as a preceding stage. However a large dynamic range of output power requires a linear wide dynamic range variable gain amplifier which is usually power consuming and hard to achieve. Also this configuration suffers from a large reduction in efficiency at lower power transmission since the standing bias current at the output stage does not scale with the output power. Also, this technique requires a very linear power amplifier for any kind of signal shaping at the input.

In non-linear power amplifiers, the input to the amplifier provides only timing information. Thus, the output power cannot be controlled through the variation in input signal amplitude as is done in linear or weakly non-linear amplifiers. Instead, output power control can be achieved effectively through a variable supply, implemented for example by a DC–DC converter. The losses in the DC–DC converter might cause the efficiency to drop to reach that of a linear power amplifier case. A new methodology based on switching a combination of power amplifiers in parallel is presented in [22] and represents an extension of the Doherty amplifier to non-linear power amplifiers.

8.3.2 Power amplifier stability issues

RF PA oscillation problems can be broadly categorised into two kinds: bias oscillation and RF oscillation. Bias oscillations that occur at very low frequencies, in the megahertz to VHF range, are caused by inappropriate or unintentional terminations at those frequencies by the bias insertion circuitry, large-value decoupling capacitors being the common cause. These oscillations have little to do with the details of the RF matching circuitry, where the RF blocking and decoupling capacitors render open circuit terminations at lower frequencies. RF oscillations, on the other hand, typically occur either in band or commonly out of band but still quite close to the desired bandwidth on the low-frequency side. These kinds of oscillations are very common in single-ended multistage designs, and their elimination will require modifications of the RF matching topology and element values. Both kinds of instability can be analysed effectively using k-factor analysis [4]. Although k-factor analysis assumes a linear two-port device, it is usually a satisfactory assumption to assume that RF oscillations in a power amplifier will more likely occur when the amplifier is backed off into its linear region, where the k-analysis is valid. In the case of deep class AB or B operation, it is necessary to increase the quiescent current to perform stability analysis with a representative amount of gain. A simple way to test stability of the PA is to run the entire circuit on a linear simulator, sweeping the frequency all the way down to DC.

Higher frequency instability will show up in a k-factor analysis of individual stages. Any single-ended design must show a k-factor greater than unity over the widest frequency sweep, extending from the low-frequency bias circuit range all the way up to the frequency at which the gain rolls off to lower than unity. Designing or modifying a circuit to obtain such a response for the k-factor typically involves some sacrifices in the in-band RF performance through use of resistive elements.

8.4   Recent progress in RF CMOS PA design

Research in the area of power amplifiers is divided into two main categories: the design and monolithic implementation of power amplifiers and the integration of linearisation techniques. While the implementation of a complete transceiver was the focus of many publications ([23–26]), the power amplifier was included in only two of the reported CMOS wireless transceivers [23, 26]. The first CMOS power amplifier that was reported targeted the 900 MHz ISM band [27] delivering output power from 20 μW to 20 mW using a 3 V supply, and implemented in 1 μm technology. The measured drain efficiency with the inductors included on-chip was 25 per cent. No input matching was included since the power amplifier was integrated in a complete transceiver [23]. The output-matching network was implemented off-chip. An extra fabrication step was used to remove the substrate beneath the inductor to improve the quality factor.

In Reference 28 a 1 W BiCMOS PA was reported. The design involves a negative resistance stage to boost the gain. The reported power added efficiency (PAE) is 30 per cent using a 5 V supply. External inductors were used as part of the interstage matching network, with the output matching network completely off-chip. Measurement results are reported for a chip-on-board die. While BiCMOS is capable of supporting other RF transceiver functions and is a strong candidate as a low-cost technology for realising a single-chip radio, the reduction in performance of a BiCMOS PA compared to GaAs PA was evident in this paper.

Many publications anticipated that CMOS would be limited only to low-power, low-performance applications. In Reference 29 a 1 W, 2.5 V supply monolithic power amplifier was reported. The PA targeted NADC standards (824–849 MHz). A gain of 25 dB was achieved through three gain stages (operating in class A, AB and C), with the output stage operating in class D (transistor is used as a switch). The power amplifier had a measured drain efficiency of 62 per cent and a PAE of 42 per cent. It did not have a high degree of integration since the output-matching network was implemented off-chip. Bond wires are also used as part of the interstage matching network.

The use of non-linear power amplifier classes has been limited to low-frequency operation until a recent publication explored the possibility of using class E power amplifiers in the 900 MHz band.

In Reference 12 a fully integrated, yet GaAs MESFET implementation of a class E PA was reported. This non-linear power amplifier outputs 250 mW at 835 MHz with power-added efficiency (PAE) of 50 per cent in a 2.5 V system. A class F amplifier was used as a driver stage to generate the required square wave input driving signal. Bias voltages were applied externally but all matching networks were included on-chip. This paper illustrated the advantages of operating at class E rather than classes C, B or F, considering the fact that it has higher optimum load and higher PAE under low-voltage operation.

Class E power amplifiers have generated wide interest after the previously mentioned publication due to the inherent high efficiency. In Reference 1 a 1.9 GHz, 1 W class E PA was implemented in 0.35 μm CMOS technology using a 2 V power supply. The input driving requirement of the output stage was greatly reduced by employing the concept of mode-locking in which the amplifier acts as an oscillator whose output is forced to run at the input frequency. The output-matching network was off-chip, and all inductors included were bond wire inductors. The measured PAE using chip on-board packaging is 48 per cent. The drawback of the mode-locking (positive feedback) technique is that the PA is prone to locking onto interfering signals picked up by the antenna from adjacent mobile users.

While the trend in most publications is to adopt non-linear power classes (class D [29], class E [1, 10, 21, 30], and class F [5, 22]) to implement high-efficiency and high-power amplifiers, the continuous decrease in the voltage breakdown of transistors for deep-submicron technologies makes the use of class E amplifiers more difficult. Class F emerges as a possible solution in this case [1, 2]. However, modern communication standards employ non-constant envelope modulation techniques that require linear power amplifiers, which means that either added linearisation circuitry would be required or traditional linear power amplifier classes are used [2, 31]. A sample of the publications listed in Table 8.2 shows that even though the inductors and capacitors that may be realised in CMOS technology are not suitable for high-performance RF circuits, CMOS transistors still have adequate gain till 2 GHz, allowing the design of low-cost hybrid 1 W amplifiers. The real merits of CMOS PAs lie in the potential for future integration. While the feasibility of a stand-alone CMOS PA does not imply its compatibility in a larger system, integration issues will

Table 8.2   Example of reported CMOS power amplifiers

Reference	Technology (μm)	Frequency (MHz)	P_{out} (dBm)	PAE (%)
27	1	900	13	30–40
29	0.8	824–849	30	42
10	0.25	900	29.5	41
1	0.35	1900	30	48
2	0.25	1950	29.2	27
31	0.35	1730	30.4	45
5	0.25	1400	24.7	43
22	0.2	900	31.7	43

rely on system, circuit and layout solutions rather than design of individual blocks. In the linearisation area, few papers have been published dealing with monolithic implementation [19, 32], while most of the published work was focused on system simulations and discrete implementations [33, 34].

In Reference 32 a phase correcting feedback system to reduce the AM to PM distortion of class E PA used in the NADC standard was presented. The system employed a limiting amplifier, a phase detector, and a phase shifter, all operating at 835 MHz. In order to reduce the phase error in the output caused by the class E amplifier, the output and input phases of the amplifier were compared and an error phase signal was generated. The error signal was applied to a phase shifter at the input of the PA. The phase correcting feedback system reduced the phase distortion from 30 degrees to 4 degrees and consumed 21.5 mW while the PA delivered 500 mW.

In Reference 19 a full monolithic CMOS implementation of the envelope elimination and restoration linearisation system that improves linearity of an efficient PA was fabricated in a 0.8 μm CMOS process. A delta modulated switching power supply was employed to extend the modulation bandwidth to fit that of the NADC. The linearisation system improves the overall efficiency from 36 to 40 per cent while increasing the maximum linear output power from 26.5 dBm to 29.5 dBm. Compared to the usual discrete implementation of EER systems used in high-power basestations, this design is amenable to integration in a low-cost CMOS technology and makes linearisation affordable to handsets.

In Reference 21 a 20 dBm power amplifier for linear amplification with non-linear components (LINC) transmitters was reported. An open loop linearized PA was realised by combining two non-linear class E amplifiers. The paper deals with a portion of the transmitter, not the whole system, and achieves 35 per cent of power added efficiency under linear operation.

8.5   Conclusions

There are several obstacles which make the implementations of a PA very difficult in CMOS technology. The use of submicron CMOS increases the difficulty of implementation due to technology limitations such as low breakdown voltage and poor transconductance. However, with the trend of lower power transmitters in the next generation, implementation of CMOS PAs with good efficiencies are becoming realistic despite steadily declining FET breakdown voltages.

8.6   References

1 TSAI, K.-C. and GRAY, P. R.: 'A 1.9 GHz 1-W CMOS class-E power amplifier for wireless communications', *IEEE Journal of Solid-State Circuits*, July 1999, **31**, pp. 962–70

2 ASBECK, P. and FALLESEN, C.: 'A power amplifier for wireless applications in a digital CMOS process'. Proceedings of the 18th Norchip Conference, Turku, Finland, Nov. 2000, pp. 28–33

3 CRIPPS, S. C.: 'RF power amplifiers for wireless communication' (Artech House, Boston, 1999)

4 GONZALEZ, G.: 'Microwave transistor amplifiers: analysis and design' (Prentice Hall, Upper Saddle River, NJ, 2nd ed., 1996)

5 KUO, T. and LUSIGNAN, B.: 'A 1.5 W class-F RF power amplifier in 0.2 μm CMOS technology', in Int. Solid-State Circuits Conf. Dig. Tech. Papers, San Francisco, USA, Feb. 2001, pp. 154–55.

6 KRAUS, H. L., BOSTIA, C. W. and RAAB, F.: Solid-state radio engineering. (John Wiley, New York, 1980)

7 GUPTA, R. and ALLSTOT, D.: 'Parasitic aware design and optimization of CMOS RF integrated circuits'. IEEE RFIC Symp, Baltimore, USA, June 1998, pp. 325–28

8 KENINGTON, P. B.: 'High-linearity RF amplifier design' (Artech House, Boston, 2000)

9 RAMAKRISHNA, S. N.: 'RF CMOS class C power amplifiers for wireless communications' Ph.D. Dissertation, University of California, Berkeley, 2001

10 YOO, C. and HUANG, Q.: 'A common-gate switched, 0.9W class E power amplifier with 41 per cent PAE in 0.2 μm CMOS', in 2000 Symposium on VLSI circuits (Honolulu, HI), June 2000, pp. 56–57

11 KUMAR, S.: 'Power amplifier linearisation using MMICs', in *Microwave Journal*, **35**, April 1992, pp. 96–104

12 SOWLATI, T., SALAMA, A., SITCH, J., RABJOHN, G. and SMITH, D.: 'Low voltage, high efficiency GaAs class E power amplifier for wireless transmitters', *IEEE J. Solid-State Circuits*, **30**, Oct. 1995, pp. 1074–80

13 SOKAL, N. and SOKAL, A.: 'Class E–a new class of high efficiency, tuned single-ended switching power amplifiers', *IEEE J. Solid-State Circuits*, June 1975, **SC-10**, pp. 168–76

14 RAAB, F. H.: 'Idealized operation of the class E tuned power amplifier', *IEEE Trans, Circuits Systems*, Dec. 1977, **CAS-24**, pp. 725–35

15 ALINIKULA, P., CHOI, K. and LONG, S.: 'Design of class E power amplifier with nonlinear parasitic output capacitance'. IEEE Trans. Circuits Systems CAS II, **46**, Feb. 1999, pp. 114–18

16 KHAN, L. R.: 'Single sideband transmission by envelope elimination and restoration'. Proceedings of IRE, July 1952, **40**

17 CHAN, K. and BATEMAN, A.: 'A 200 MHz IF BiCMOS chip for linear LINC transmitters'. IEEE Trans. Circuits Systems CAS I, June 1995, **42**, pp. 321–33

18 RAZAVI, B.: RF microelectronics (Prentice-Hall, Upper Saddle River, NJ, 1998)

19 SU, D. and MCFARLAND, W.: 'An IC for linearizing RF power amplifiers using envelope-elimination and restoration', *IEEE J. Solid-State Circuits*, Dec. 1998, **33**, pp. 2252–58

20 SHI, B. and SUNDSTROM, L.: 'A low stress 20 dBm power amplifier for LINC transmission with 50% peak PAE in 0.25 μm CMOS'. ESSIRC Dig. Tech. Papers, Duisburg, Germany, Sept. 1999, pp. 282–85

21 TARSIA, M., KHOURY, J. and BOCCUZZI, V.: 'A low stress 20 dBm power amplifier for LINC transmission with 50 peak PAE in 0.25 mm CMOS', in ESSIRC Dig. Tech. Papers, Sept. 2000, pp. 100–103

22 SHIRVANI, A., SU, D. and WOOLEY, B.: 'A CMOS RF power amplifier with parallel amplification for efficient power control', in In Int. Solid-State Circuits Conf. Dig. Tech. Papers, San Francisco, USA, Feb. 2001, pp. 156–57

23 ROFOUGARAN, A., CHANG, G., RAEL, J., CHAN, J. Y. C., ROFOUGARAN, M., CHAN, P. J., DJAFARI, M., KU, M.-K., ROTH, E. W., ABIDI, A. A. and SAMUEL, H.: 'A single-chip 900 MHz spread spectrum wireless transceiver in 1 mm CMOS part I: architecture and transmitter design', *IEEE J. Solid-State Circuits*, April 1998, **33**, pp. 513–34

24 RUDELL, J., OU, J.-J., CHO, T. B., CHIEN, G., BRIANTI, F., WELDON, J. A. and GRAY, P. R.: 'A 1.9 GHz wide band IF double conversion CMOS receiver for cordless telephone applications', *IEEE Journal of Solid-State Circuits*, **32**, Dec. 1997, pp. 2071–88

25 STEYAERT, M., BORREMANS, M., JANSSENS, J., et al.: 'A 900 MHz/1.8-GHz CMOS receiver for dual-band applications', in In Int. Solid-State Circuits Conf. Dig. Tech. Papers (San Francisco), Feb. 1998, pp. 48–49

26 MELLY, T., PORRET, S., ENZ, C. C. and VITTOZ, E.: 'A 1.2 V, 433 MHz, 10 dBm, 38% global efficiency FSK transmitter integrated in a standard digital CMOS process'. Proceedings of Custom Integrated Circuits Conference (Orlando, Fl.), May 2000, pp. 179–82

27 ROFOUGARAN, M., ROFOUGARAN, A., OLGAARD, C. and ABIDI, A.: 'A 900 MHz CMOS RF power amplifier with programmable output power', in IEEE Symp. VLSI Circuits Dig. Tech. Papers (Honolulu, HI.), June 1994, pp. 133–34

28 WONG, S., BHIMNATHWALA, H., LUO, S., HALAI, B. and NAVID, S.: 'A 1W 830 MHz monolithic BiCMOS power amplifier', in *In Int. Solid-State Circuits Conf. Dig. Tech. Papers*, San Francisco, USA, Feb. 1996, pp. 52–53

29 SU, D. and MCFARLAND, W.: 'A 2.5-V, 1-Watt monolithic CMOS RF power amplifier', in Proceedings of Custom Integrated Circuits Conference, Santa Clara, CA, May 1997, pp. 179–82

30 MERTENS, K., STEYAERT, M. and NAUWELAERS, B.: 'A 700 MHz 1 W fully differential class E power amplifier in CMOS', in *ESSIRC Dig. Tech. Papers* (Stockholm), Sept. 2000, pp. 104–07

31 FALLESEN, C. and ASBECK, P.: 'A highly integrated 1W CMOS power amplifier for GSM-1800 with 45% PAE', in Proceedings of the 18th Norchip Conference (Turku, Finland), Nov. 2000, pp. 140–45

32 SOWLATI, T., GRESHISHCHEV, Y. and SALAMA, A.: 'Phase-correcting feedback system for class E power amplifier', *IEEE J. Solid-State Circuits*, April 1997, **32**, pp. 544–50

33 REY, C. G.: 'Predistorter linearizes CDMA power amplifier', *Microwave and RF Magazine*, Oct. 1998, **44**, pp. 114–23

34 SUN, J., LI, B. and CHIA, M. W.: 'Linearized and highly efficient CDMA power amplifier,' *Electronic Letters*, May 1999, **35**, pp. 786–87

Chapter 9

Parasitic-aware RF IC design and optimisation

Kiyong Choi, Jinho Park and David J. Allstot

9.1 Introduction

Based on the principles of radio communication theorised by Maxwell in 1862, the wireless communications industry has grown exponentially during the past few decades. Nowadays, almost everyone owns at least one wireless communications product such as a television, radio, cellular phone, cordless phone, pager, wireless modem, wireless computer peripheral, etc. Because of the burgeoning popularity of wireless communications, research is ongoing to develop small-size, lightweight, low-cost, and highly power-efficient products. Moreover, consumers continually demand less expensive and more portable personal communications devices with ever increasing functionality. Consequently, intense worldwide research is focused on the design of RF communication circuits that can be integrated with both analogue and digital subcircuits in CMOS system-on-chip solutions.

Two major constraints in the design and optimisation of RF frontend circuits in fine-line CMOS technology are active devices that are inferior to their GaAs and SiGe counterparts, and low-quality parasitic-laden passive components owing to the use of lossy silicon substrates. To overcome these drawbacks, the *parasitic-aware* synthesis paradigm has been developed. Simulated annealing is one of the key algorithms used in the parasitic-aware CAD design and optimisation methodology.

Unlike in baseband circuits, the parasitic effects are severe in high-frequency circuits. In the case of baseband circuit design, minimising parasitics is usually sufficient. However, the parasitics must be carefully modelled and considered as *part of the design process* in high-frequency circuit synthesis. For example, if an RF circuit is first designed without considering parasitics, simulations show that it often loses most of its performance when parasitics are subsequently included. On the other hand, if parasitics are considered as part of the design, it is difficult to find an analytic solution by hand. The parasitic-aware synthesis paradigm has been developed to address these difficulties.

One of the burning issues in optimisation is how best to consider process, voltage and temperature (PVT) variations. Circuit simulation with PVT variations is very time consuming because Monte Carlo methods are normally required, and it is difficult to make them part of the optimisation process. Consequently, the analysis of PVT variations as part of a post-optimisation process will be discussed.

In this chapter, parasitic-laden inductor modelling is first discussed. The conventional simulated annealing algorithm is then described and compared to classical gradient descent optimisation techniques. Finally, the shortcomings of the conventional simulated annealing algorithm are addressed. An innovative adaptive simulated annealing algorithm is proposed to improve upon the conventional algorithm. Post-optimisation PVT variation analysis is discussed, and two high-frequency circuit examples show that parasitic-aware CAD optimisation is essential in RF circuit design.

Before discussing the CAD optimisation techniques, compact inductor models are briefly reviewed in order to understand the parasitics associated with on-chip spiral inductors.

9.2   On-chip inductor modelling

Inductors are an essential component of RF circuit design and are used in impedance matching networks, resonant networks, etc. Accurate compact modelling of on-chip inductors is very important because their parasitics can significantly degrade the performance of high-frequency circuits. Moreover, analytical representations of the parasitic components are needed in order to consider them as part of the parasitic-aware design and optimisation processes. The two types of inductors used in RF integrated circuit design are on-chip spirals and bond-wire inductors.

9.2.1   *Spiral inductors*

Monolithic spiral inductors play a key role in increasing the integration level of RF chips. However, their performance is far inferior to discrete off-chip inductors because of their large series and substrate loss resistances. Figure 9.1 shows the cross-sectional and top views of a square spiral inductor. Two types of parameters are needed to model the spiral inductor, namely, process-and design-controlled parameters. Process-controlled parameters (Figure 9.1a) include oxide thickness, metal thickness, substrate resistivity, etc., while design-controlled parameters (Figure 9.1b) include metal width, number of turns, centre spacing, metal line spacing, etc. The relationships between the parasitic component values and the design and process parameters are complex and difficult to model accurately. For example, to reduce the series and substrate resistances of the spiral inductor and increase its quality factor (Q), the metal width can be increased. However, this increases the oxide capacitance between the spiral and the substrate, thus reducing the self-resonance frequency (F_r).

There are several ways to find the parasitic component values associated with an inductor. One method involves the development of approximate analytic equations [1],

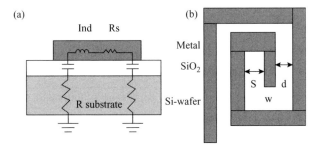

Figure 9.1   *(a) Cross-sectional and (b) top views of a parasitic-laden monolithic square spiral inductor*

and another uses electromagnetic (EM) simulations. A purely empirical method involves first fabricating and characterising *all* inductors, in the chosen CMOS technology, before they are used in subsequent circuit designs. The first method is easy to use and moderately accurate to frequencies of about 2 GHz. However, at higher frequencies, the compact model component values are insufficiently accurate due to various high-frequency effects such as the skin effect, etc. The EM simulation method can be more accurate for higher frequencies depending on knowledge of the CMOS process details, the complexity and type of the compact model, and the capabilities of the simulator. However, EM simulations require very long simulation times and are often not sufficiently accurate. The full-blown empirical characterisation method provides good models for all frequencies, but it is expensive and time consuming to fabricate inductors prior to the design and optimisation processes.

An alternative approach for modelling spiral inductors combines aspects of EM simulation and full-blown characterisation. We first design, fabricate and characterise three spirals with identical geometric features spanning a range of values. (We call this the '*three bears*' approach because the smallest (baby), a mid-size (mother), and the largest (father) inductor values provide sufficient data.) After fabrication and characterisation, the results are used to *calibrate* an EM simulator to fit the measured data. It can now accurately predict parasitics for interpolated inductance values.

Figure 9.2 shows a typical measured frequency response of an on-chip spiral inductor. In 0.35 μm CMOS, a metal-3 spiral with 6.25 turns (9 nH, metal width = 15 μm, centre spacing = 101.4 μm, and metal spacing = 1.2 μm) exhibits a peak Q of 4 at 3.2 GHz and a self-resonant frequency of 3.2 GHz. Parasitic-aware optimisation requires that the parasitic values be expressed versus inductance (Figure 9.3). The model equations are obtained using the three bears modelling approach, EM simulations for interpolation, and the `polyfit` command in MATLAB.

9.2.2   Bond wire inductor

Another possible on-chip inductor is the bond wire, which has several advantages over the spiral inductor. Since the bond wire is often gold, rather than aluminium, and its radius is large (about 30 μm) compared to the dimensions of the spiral, its

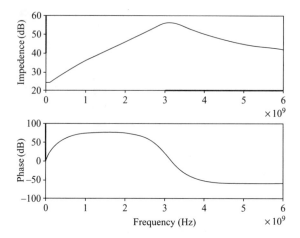

Figure 9.2   Typical measured frequency response of an on-chip square spiral inductor

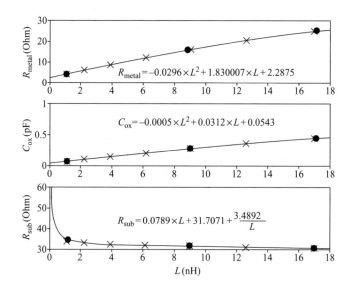

Figure 9.3   Parasitic values (see Figure 9.1) versus inductance for on-chip square spiral inductors. The dark circles represent values extracted from measurements using the 'three bears' modelling approach. The 'x' values are obtained using a calibrated EM simulator

series resistance is much smaller, and therefore, its quality factor is also much higher. Figure 9.4 shows a bond wire inductor with its important parasitics. Note that the bonding pads contribute C_{ox} and R_{sub} which are constant with respect to the inductance value (i.e. the length of the bond wire).

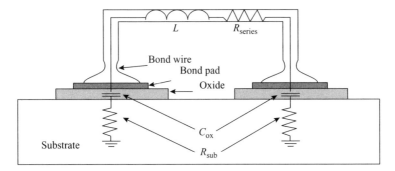

Figure 9.4   A typical bond wire inductor with parasitics

As a rule of thumb, each 1 mm length of the bond wire contributes 1 nH of inductance. A first-order formula commonly used to calculate the bond wire inductance value is [2]

$$L = \frac{l}{5}\left(\ln\left(2\frac{l}{r}\right) - 0.75 + \frac{r}{l}\right) \tag{9.1}$$

where L is the inductance in nH, l is the length of the bond-wire in mm, and r is its radius in mm. Parasitic component values for a typical bond pad are $C_{ox} = 200$ aF and $R_{sub} = 20\ \Omega$.

The series resistance is

$$R_{series} = \frac{\rho l}{A} \tag{9.2}$$

where ρ is the resistivity of gold, 2.35 μohm-cm, l is the length of the wire in cm, and A is its cross-sectional area in cm^2. For use at high frequencies, the skin effect that increases the effective resistivity at high frequencies must be considered. The associated skin depth δ is

$$\delta = \sqrt{\frac{2\rho}{\omega\mu}} \tag{9.3}$$

Since the permeability (μ) of gold is 1.26 μF/m, the skin depth is 2.57 μm at 900 MHz, for example. Assuming a typical radius for the bond wire inductor of 30 μm, δ is clearly much smaller than r; hence, the approximate skin area of conduction is

$$A_{eff} = 2\pi r\delta \tag{9.4}$$

Using (9.3) and (9.4), the series resistance is 0.05 ohm per mm.

Traditionally, the manufacturability and repeatability of bond wire inductors have been concerns. However, it has recently been shown that machine-bonded wires of the type used in manufacturing exhibit less than ±5 per cent inductance variations and less than ±6 per cent Q variations [3].

9.3   Parasitic-aware optimisation

As intimated above, it becomes virtually impossible to find tractable analytic solutions for RF circuits when all of the parasitic components are considered. Hence, we employ *parasitic-aware* numerical CAD design and optimisation techniques to find solutions. Conventional gradient descent optimisation techniques are ubiquitous. However, they are not well suited for RF circuit synthesis because they do not possess a probabilistic hill climbing capability, and can therefore easily become trapped in suboptimal local minima in the complex design space. To illustrate this effect, assume the initial design point is in Region I or III (Figure 9.5). Conventional gradient descent methods (Figure 9.6) would find a local minimum at A or C, not the global minimum solution at B. Hence, optimisation methods are required that can find the global minimum for arbitrary initial design points; one such global search heuristic is *simulated annealing*.

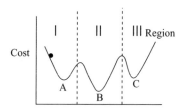

Figure 9.5   Local minima (A and C) and a global minimum (B) for a simple cost function

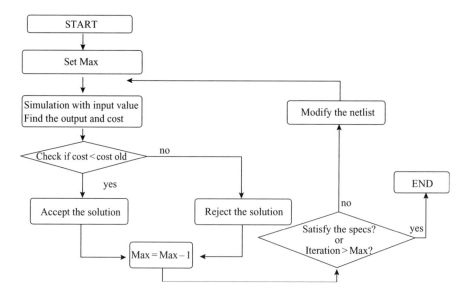

Figure 9.6   Typical flow chart for gradient descent optimisation

9.3.1   Simulated annealing

As the name suggests, the simulated annealing algorithm originated from metal annealing technology. In metal annealing, a product such as a knife is heated to a high temperature during manufacture, and then cooled at an optimum rate so that its total number of defects is minimized. After annealing, the positions of the metal atoms are stable and the bonds between them are strong; in other words, a minimum energy system is created. Simulated annealing mimics this behaviour [4–6].

Simulated annealing provides a statistical hill climbing capability, and thus avoids, with high probability, being trapped in undesired local minima. However, the hill climbing process needs to be controlled to minimise simulation time. In the simulated annealing technique, two controls on the hill climbing process are the slope of the cost function and the Temp coefficient:

$$\text{Rand } (0, 1) < \exp\left(\frac{\text{oldcost} - \text{cost}}{\text{Temp}}\right). \tag{9.5}$$

The left side of (9.5) is a random number between 0 and 1, and because of the preceding conditional operation shown in Figure 9.7, oldcost − cost ($-\Delta$cost) is always less than 0. Therefore, the right side of (9.5) is also always less than 1. If Temp is high and the right side of (9.5) is close to 1, the probability of hill climbing is high. On the other hand, if Δcost is large, the right side of (9.5) approaches 0, and the probability

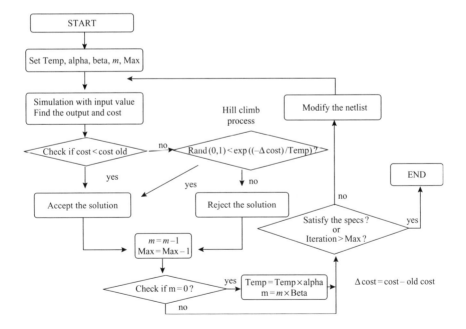

Figure 9.7   Flow chart of the parasitic-aware simulated annealing optimisation methodology

Table 9.1   Relationships between probability of hill climbing and parameters Temp and oldcost-cost ($\Delta cost$)

Probability of hill climb	Temp	Δ(cost)
High	High	Low
Low	Low	High

of hill climbing is low. Table 9.1 summarizes the general relationships between the probability of hill climbing and the coefficients Temp and Δcost.

In an implementation of the simulated annealing algorithm, the parameter Temp is, in turn, controlled by the parameter alpha. Initially, Temp is set to a high value that decreases during optimisation according to the value of alpha. Figure 9.7 shows a typical flow chart of simulated annealing optimisation; it requires the setting of five parameters: alpha, beta, Temp, *m* and Max. Temp is the temperature parameter required in equation (9.5), and alpha is used in controlling the value of Temp. *m* is the number of iterations for the same Temp value, and beta is used for controlling the *m* factor. Max is the maximum number of iterations allowed.

In the gradient descent search optimisation method shown in Figure 9.6, only the slope of the cost function determines if a solution is accepted. In contrast, simulated annealing uses an iterative loop with an optimum temperature-cooling algorithm. Hence, there are more chances of finding an acceptable solution according to equation (9.5).

Figure 9.8 shows the computational flow details of the simulated annealing approach. The simulated annealing code requires a SPICE Netlist, and a setup file that includes the design parameters that are to be optimised, the cost function information, and the parameters described above including Temp, alpha, Max, etc. First, the setup file is used to modify the SPICE Netlist according to the assumed initial solution; then, SPICE is run in order to evaluate the given cost function. After evaluating it, the simulated annealing optimiser conditionally accepts the solution according to the five key parameters. Accepted solutions are saved to the .m file for later analysis and plotting using MATLAB.

A parasitic-aware RF CAD design and optimisation tool should exhibit two key features. First, it should find acceptable solutions efficiently in terms of the required number of computer cycles. Second, it should possess the ability to escape local minima in the complex design space in order to maximise the probability of finding the global minimum. While the general simulated annealing algorithm offers a hill climbing capability that usually satisfies the second objective, it has several significant shortcomings for parasitic-aware RF circuit synthesis. A critical parameter in determining the cooling schedule in conventional simulated annealing is the Temp

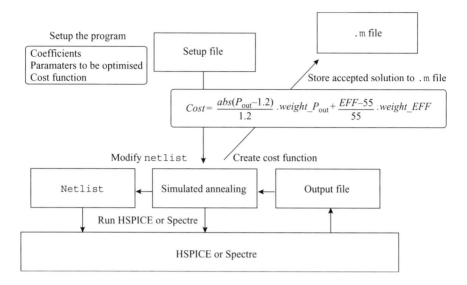

Figure 9.8   Flow diagram of the inner loop of the parasitic-aware simulated annealing algorithm

coefficient. Not only is it difficult and time consuming to determine an optimum value for this key empirical parameter, a prohibitive number of iterations is often required to find a suitable set of acceptable solutions once the value of Temp is set.

In this chapter, a simulated annealing algorithm is described that features adaptive determination of the Temp coefficient, and a *tunnelling* process to speed the global and local searches, respectively. Hence, the computer time required for parasitic-aware RF circuit synthesis is significantly reduced and the probability of finding the global minimum is substantially increased, compared to previous simulated annealing approaches.

9.3.2   Adaptive simulated annealing

The simulated annealing parasitic-aware synthesis approach has three major component parts: the tunnelling process, the local optimisation technique, and the adaptive Temp coefficient method.

9.3.2.1   Tunnelling process

A well-known advantage of simulated annealing is its hill climbing ability, which is determined by the Temp coefficient in the cooling schedule and the slope of the hill. Showing a simple cost function that depends on a single design variable, X, Figure 9.9a depicts hill climbing in conventional simulated annealing. There is a finite non-zero probability of escaping local minima in search of the global minimum. However, if Temp is fixed and the hill is steep, an unacceptably large number of iterations may be required to climb from point A to point B, even though there are

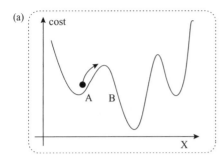 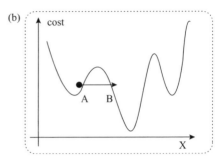

*Figure 9.9   (a) Hill climbing in simulated annealing, and (b) the tunnelling pro-
cess in adaptive simulated annealing for a cost function versus design
variable, X*

no better solutions between those two points as indicated by the higher values of the
cost function.

In the case of a genetic optimisation algorithm, this drawback is often overcome
using TABU search methods [7]. With TABU, unacceptable points are recorded and
not simulated again so the total number of iterations to find an acceptable set of
solutions is reduced. However, the TABU method is not reliably applied with the
simulated annealing algorithm because it effectively inhibits the hill climbing process. That is, the solution search may become stuck at a local minimum because
TABU disallows hill climbing over known unacceptable solutions in search of the
optimum.

To achieve the advantages of TABU while avoiding its catastrophic disadvantage
in simulated annealing as described above, a new *tunnelling* technique is developed, as
illustrated conceptually in Figure 9.9b. Rather than wasting time hill climbing through
a region of known unacceptable solutions, or getting stuck, the optimiser simply senses
the beginning of a climb and enables *tunnelling* from point A to point B. Whereas
TABU considers unacceptable solutions only once, the tunnelling method *never* considers many unacceptable solutions. Hence, the required number of iterations needed
in parasitic-aware synthesis is greatly reduced.

Figures 9.10 compares conventional simulated annealing and adaptive simulated
annealing with tunnelling. The curves represent a cost function versus values of a
single design variable, X. Note that the cost function has several local minima, and
that the 'o' and '×' points portray accepted and unaccepted solutions, respectively. As
illustrated in the example of Figure 9.10a, conventional simulated annealing wastes
iterations finding unacceptable solutions between points A and B, far above the global
minimum (lowest cost) solution. In general, only a small percentage of the solutions
are the desired low cost ones.

In contrast, when the tunnelling process is applied as depicted in Figure 9.10b,
no acceptable and few unacceptable solutions are considered between points A and
B. That is, the searchable design space is reduced because many unacceptable points

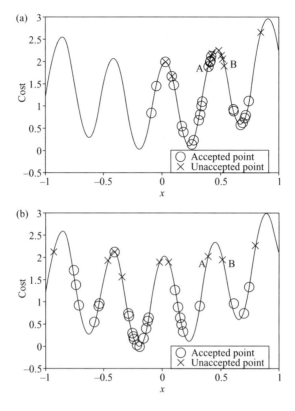

Figure 9.10 *(a) Conventional simulated annealing, and (b) adaptive simulated annealing results for a cost function versus a design variable, x*

between A and B are not simulated. Hence, the new tunnelling technique focuses on more desirable lower cost acceptable solutions. Thus, there is a higher probability of finding the global minimum quickly without becoming trapped in a local minimum.

The initial version of the tunnelling algorithm uses two key parameters called the *tunnelling threshold* and the *tunnelling radius*. The tunnelling threshold determines the conditions when tunnelling is invoked, while the tunnelling radius specifies the maximum distance a simulation point tunnels in the design space. The tunnelling radius is a critical coefficient in the tunnelling process. If it is too small, the probability of tunnelling through a hill is low; if it is too large, the solution might tunnel so far past the optimum solution that it may never be found. A simple method of determining the tunnelling radius is described below.

Figure 9.11 shows a flow chart of the tunnelling process. As indicated, when tunnelling occurs, a fraction of the tunnelling radius is added to the previous parameter values to determine the new design variable values.

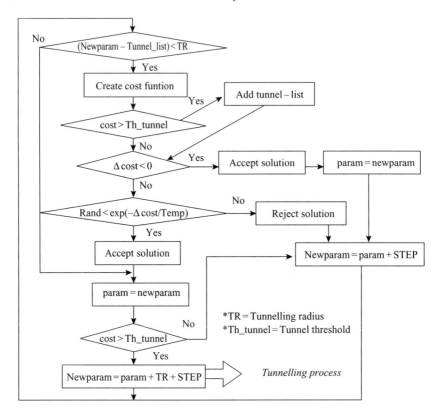

Figure 9.11   Flow chart of simulated annealing with tunnelling process

9.3.2.2   Local optimisation algorithm

One major reason that a conventional simulated annealing optimiser requires a large number of iterations to find an acceptable solution is that it chooses its next simulation point randomly. For example, if the optimisation problem involves 20 design variables, and each is randomly decreased or increased, then there are $2^{20} = 10^6$ possible directions for the next simulation point! The low probability of choosing a good direction leads to a large number of iterations to find an acceptable solution. On the other hand, if the new direction is set using only information from the previous simulation point, it is easy to become stuck in a local minimum.

As mentioned earlier, the key equation of conventional simulated annealing repeated here is

$$\text{Rand}\,(0,\,1) < \exp\left(\frac{\text{oldcost} - \text{cost}}{\text{Temp}}\right). \tag{9.5}$$

If the parameter Temp is large, the probability of hill climbing is high, and vice versa.

Figure 9.12 depicts the new method of determination of the direction to the next simulation point based on information from both the present and previous points. The

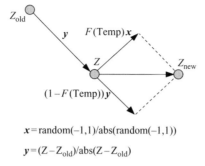

$x = \mathrm{random}(-1,1)/\mathrm{abs}(\mathrm{random}(-1,1))$

$y = (Z - Z_{\mathrm{old}})/\mathrm{abs}(Z - Z_{\mathrm{old}})$

Figure 9.12   Weighted vector sum of previous and random directions for determining the direction to the next simulation point

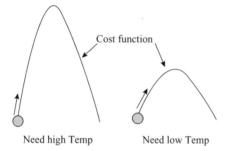

Figure 9.13   The relation between slope of the cost function and Temp coefficient

vector y is defined between the previous and present points, the vector x is chosen randomly, and the new direction is a weighted sum of the two. If Temp is high, the new direction is predominantly random, but if Temp low, the new direction is almost identical to y. Hence, the next simulation point Z_{new} is

$$Z_{\mathrm{new}} = Z + (F(\mathrm{Temp}) \cdot x + (1 - F(\mathrm{Temp})) \cdot y) \cdot \mathrm{Rand}\,(0, 1) \cdot \mathrm{Stepsize}. \quad (9.6)$$

In (9.6), $F(\mathrm{Temp})$ is a function that is proportional to the Temp coefficient.

9.3.2.3   Adaptive Temp coefficient determination

In conventional simulated annealing, the value of the temperature parameter, Temp, is critical in minimising the number of iterations required to find acceptable solutions. Figure 9.13 symbolises the relationship between Temp and the cost function slope. If Temp is too large, global search dominates and iterations are wasted. On the other hand, if Temp is too small, local search dominates and the optimiser gets stuck in local minima. Experience shows that the optimum value of Temp is strongly dependent on the RF circuit topology; i.e. high Temp is required for circuits with steep cost function

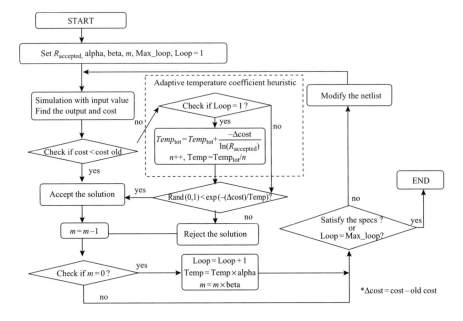

Figure 9.14   The block diagram of adaptive temperature coefficient heuristic

and vice versa [8]. Greater computational efficiency is achieved by adapting Temp using an estimate of the cost function slope. An implicit relationship between Temp and the cost function slope estimate is shown in (9.7a) below. The adaptive value of Temp is obtained by replacing the random number with a fixed $R_{accepted}$, solving the equation for Temp, and averaging as shown in (9.7b).

$$\text{Rand}\,(0,1) < \exp\left(\frac{\text{oldcost} - \text{cost}}{\text{Temp}}\right) \qquad (9.7a)$$

$$\text{Temp} = \text{AVG}\left(\frac{\text{oldcost} - \text{cost}}{\ln(R_{accepted})}\right). \qquad (9.7b)$$

Figure 9.14 illustrates the computational flow for adaptively determining Temp. For a given circuit topology, experience shows that the estimate of Temp obtained from m iterations within the first optimisation loop (Loop $= 1$) provides sufficient accuracy for further optimisation cycles.

Specifically, all positive cost function differences between iteration points within the first loop are weighted and averaged to estimate Temp. (Of course, the algorithm allows Temp to be estimated from averages over additional loops, if desired.)

While the adaptive Temp algorithm eliminates the need to specify an initial value for Temp, a new empirical parameter, $R_{accepted}$, is introduced. It represents the initial probability of hill climbing and ranges from 0.7 to 0.9 depending on the desired optimisation speed. i.e. it determines the balance between global and local search capabilities. Fortunately, a significant advantage accrues from this approach; $R_{accepted}$ is not dependent on the circuit topology like Temp.

9.3.3   Post-PVT variation optimisation

A key indicator of the robustness of a design is the sensitivity to process, voltage and temperature (PVT) variations. PVT sensitivity should be considered in conjunction with the optimisation algorithm. One approach involves specifying PVT variations as part of the cost function to be optimised. Unfortunately, since the evaluation of PVT variations involves many SPICE simulations at each possible solution point, correspondingly more iterations are required to find acceptable solutions. For example, if 200 circuit simulations are required to obtain PVT sensitivity information at each design point, and 10,000 iterations are required to find an optimum solution before considering PVT variations, the total iteration count grows to $200 \times 10,000 = 2,000,000$. The problem here is that PVT variations are being considered for the majority of solutions that are unacceptable to begin with.

A new method for considering PVT sensitivity is *post-optimisation PVT analysis* wherein PVT variations are simulated for only the accepted solutions. After running PVT Monte Carlo simulations, the average and standard deviation of the resulting cost functions are evaluated and that information is used to possibly redefine an optimum solution. With this approach, the optimum solution may differ from that obtained before *post-optimisation PVT analysis*.

If, after post-optimisation PVT simulations, the best solution does not meet the specifications, the previously accepted solutions are added to the tunnelling list with the previously unaccepted solutions so that they are not considered further. Figure 9.15 shows the block diagram of the post-optimisation PVT simulation strategy. As indicated, if the optimum solution does not satisfy the specifications after the first optimisation run, the adaptive simulated annealing algorithm is run again. However, note that the simulated annealing algorithm will run much faster during the second optimisation run because of the tunnelling list; it effectively shrinks the design space for each subsequent simulation run by not resimulating known bad solutions.

9.3.4   Simulation results

A test cost function is chosen as

$$\text{Cost} = \cos(14.5x - 0.3) + (x + 0.2)x + 1. \tag{9.8}$$

Figure 9.10 displays the test cost function versus values of the single design variable, x. It exhibits several local minima and a global minimum. Figures 9.16–9.18 show comparative results between conventional simulated annealing and the new adaptive simulated annealing with tunnelling. The total number of iterations is defined as

$$\text{Total\_iteration} = \text{Maxloop} \times m. \tag{9.9}$$

The maximum number of external loop operations, Maxloop, affects inversely the value of the Temp coefficient. That is, if Maxloop is increased, Temp is decreased.

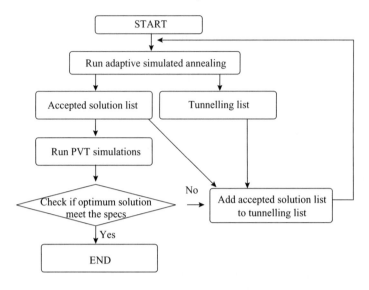

Figure 9.15   Block diagram of post-optimisation PVT variation analysis

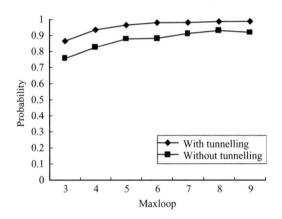

Figure 9.16   Probability of finding the global minimum versus Maxloop for conventional versus adaptive simulated annealing

m is the maximum number of iterations with the same value of Temp within a given external computation loop. As shown in Figures 9.16 and 9.17, the adaptive annealing algorithm has a higher probability of finding a global minimum given a fixed maximum number of iterations.

   Tunnelling employs a coefficient called the *tunnelling radius* that determines how far the simulation point jumps when tunnelling occurs. This is a critical parameter as shown in Figure 9.18; if the radius is too small, tunnelling occurs infrequently with results similar to conventional simulated annealing. On the other hand, if the radius

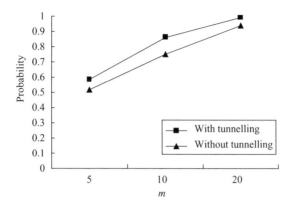

Figure 9.17   Probability of finding the global minimum versus m

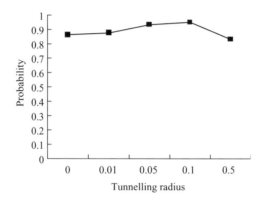

Figure 9.18   Probability of finding the global versus tunnelling radius for conventional versus adaptive simulated annealing

is too large, unaccepted solutions might obscure the optimum point so that the global minimum is seldom found. Since PVT variations in an IC range over ±20 per cent, a value of 5–10 per cent of the parameter value for the *tunnelling radius* is somewhat arbitrarily chosen.

Tunnelling applied with a simple cost function is shown to be more effective than the conventional hill climbing process in finding the global minimum with a limited number of iterations. Next, tunnelling is applied to the synthesis of an RF power amplifier in fine-line CMOS technology.

9.4   RF circuit syntheses

In the remainder of this chapter, two different types of amplifiers are optimised comparing the conventional and new adaptive simulated annealing with tunnelling algorithms: a class-E switching CMOS RF power amplifier and a CMOS RF wide-band distributed amplifier. Basic operation of the two amplifiers is explained before the circuits are synthesised using the parasitic-aware paradigm.

9.4.1   Class-E power amplifier

The class-E power amplifier, invented by Sokal in 1975, uses a transistor as a switch [9, 10]. If the switch is ideal, the voltage drop across its terminals is zero when it is on. On the other hand, when the switch is off, the current through it is zero. Hence, the power dissipated in the switch is zero since it is just the product of the voltage and current, one of which is always zero. However, if the switch is not ideal, there exist overlap times between the non-zero current and non-zero voltage waveforms which create power losses that reduce the power efficiency. The basic class-E concept is to use time delays implemented using capacitors and inductors to minimise the overlap times between the current and voltage waveforms in order to regain high efficiency. Figure 9.19 shows the basic structure of a class-E power amplifier and the associated waveforms. Since it uses a transistor as a switch, it is a non-linear power amplifier and is best suited for non-linear digital modulation schemes such as GMSK, MSK, etc.

Figure 9.20 shows a multistage power amplifier block diagram with typical gain and drain efficiency (η) distributions. To obtain high power gain, a three-stage PA is used. Figure 9.20a shows an example achieving 30 dB gain with 50 per cent drain efficiency. The drain efficiency of the third stage is most important because of its large output power. Hence, as shown in Figure 9.20b, the last two stages are class-E for high drain efficiency while the first stage is class-AB for high gain.

The first two stages use spiral inductors while the third stage and all choke inductors use bond wires. High quality bond wires must be used in the third stage because its output power is large.

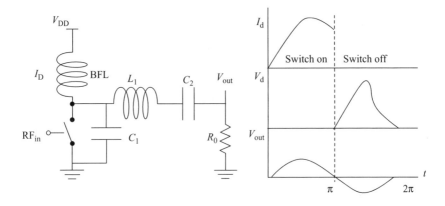

Figure 9.19   Basic circuit topology of a class-E power amplifier and its associated waveforms

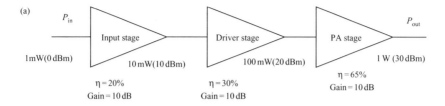

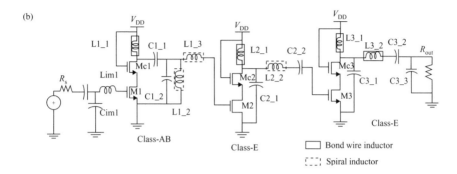

Figure 9.20   *Three-stage CMOS power amplifier (class-AB, class-E, class-E) showing gain and drain efficiency (η) distribution*

To achieve one watt of output power

$$P_{\text{out}} = 0.577 \frac{V_{\text{DD}}^2}{R_{\text{eff}}} \tag{9.10}$$

$$R_{\text{eff}} = 0.577 \frac{V_{\text{DD}}^2}{P_{\text{out}}} \tag{9.11}$$

The resistance seen looking into the matching network connected to the 50 Ω antenna needs to be about 5 Ω. Unfortunately, the parasitic series resistance of the spiral inductor is greater than 5 Ω, and since it forms a voltage divider with the output load, it is impossible to deliver 1 W output power to the antenna. Hence, the bond wire inductor with its extremely small series resistance is used in the third stage.

To reduce the voltage stresses on the MOSFETs, cascodes are used in all three stages. HSPICE simulations were performed using the TSMC 0.35 μm CMOS parameters. The most important cost function components of a switched power amplifier are output power and drain efficiency.

Optimisation of this particular RF power amplifier involves 22 design parameters; the cost function is aimed at achieving at least 55 per cent drain efficiency with at least 1.2 W output power.

The specific cost function of the power amplifier examined here is

$$\text{Cost} = \frac{\text{abs}(P_{\text{out}} - 1.2)}{1.2} \cdot \text{weight\_}P_{\text{out}} + \text{abs} \frac{(\text{EFF} - 55)}{55} \cdot \text{weight\_EFF.} \tag{9.12}$$

where weight_P_{out} and weight_EFF are weight parameters of P_{out} and EFF (efficiency).

According to equation (9.12), the output power needs to be exactly 1.2 watt, and the drain efficiency needs to be greater than 55 per cent to minimise the cost function. In this simulation, weight_P_{out} and weight_EFF are set to 50.

9.4.2   Class-E power amplifier optimisation results

Since the value of Temp is critical for both simulated annealing algorithms in comparing the tunnelling and conventional hill climbing processes, the new adaptive Temp coefficient algorithm is used to set Temp equally in both, for comparison purposes. The maximum iteration count is set to 20,000 in both cases.

Figure 9.21 compares the two approaches for a total of 12 synthesis runs. With tunnelling, the optimum is always found within 8,000 iterations; 41 per cent of the simulations find the optimum within 4,000 iterations. In contrast, for conventional simulated annealing without tunnelling, 25 per cent of the simulation runs do not find the optimum solution within 20,000 iterations; moreover, only 20 per cent of the runs find the optimum within 4,000 iterations. Clearly, the tunnelling process is superior to conventional hill climbing in escaping from local minima and accelerating the optimisation.

Figure 9.22 plots the cost function trajectories versus the number of iterations for both the conventional and adaptive simulations. As shown, when the cost is low, many simulations are wasted in conventional annealing compared to the advanced local optimisation algorithm.

The tunnelling process is superior to hill climbing in escaping from local minima, and the local optimisation algorithm achieves faster convergence to the optimum solution when Temp is low.

Table 9.2 summarises the 22 design parameter values shown in Figure 9.20 before and after optimisation. As indicated, most of the 22 design values change significantly

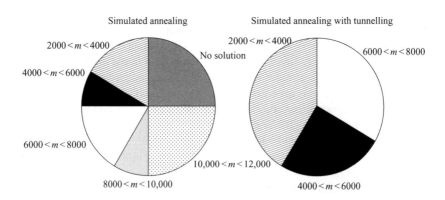

Figure 9.21   *The number of iterations required to find the optimum solution for simulated annealing without and with tunnelling (12 synthesis runs) with power amplifier*

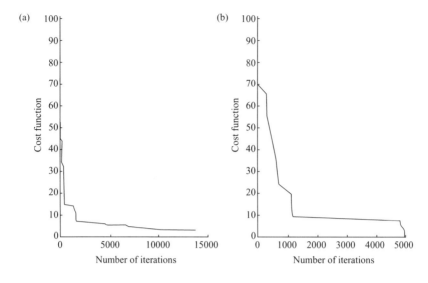

Figure 9.22   Cost function trajectory versus the number of iterations in (a) conventional and (b) adaptive simulated annealing with tunnelling. Note the different x-axis scales

Table 9.2   Summary of the 22 design parameter values before and after optimisation (see Figure 9.20)

Value	Ideal	After optimisation	Value	Ideal	After optimisation
Cim1	7.5 pF	8.6 pF	C2_1	6.6 pF	0.68 p
Lim1	5.5 nH	6.8 nH	C2_2	6.6 pF	0.67 p
Bias1	1 V	0.7 V	L2_1	10 nH	3.9 n
M1,MC1	2 mm	1.25 mm	L2_2	9.51 nH	3.8 n
C1_1	20 pF	11 pF	Bias3	1 V	1.3 V
C1_2	1.08 pF	0.48 pF	M3,Mc3	15 mm	12.4 mm
L1_1	20 nH	8.7 nH	C3_1	5.3 pF	0.14 p
L1_2	5 nH	16 nH	C3_2	14 pF	14.1 p
L1_3	2.9 nH	2.85 nH	C3_3	14 p	7.9 p
Bias2	1 V	1 V	L3_1	11.7 n	1.54 n
M2,Mc2	4 mm	3.1 mm	L3_2	5 n	4.7 n

during the optimisation process. In particular, RF chokes L1_1, and L2_1, and L3_1 become smaller than in the ideal case to reduce their series loss resistances.

The desirability of parasitic-aware synthesis is vividly illustrated in Figure 9.23 and Table 9.3. With ideal parasitic-free inductors on all three stages, the output

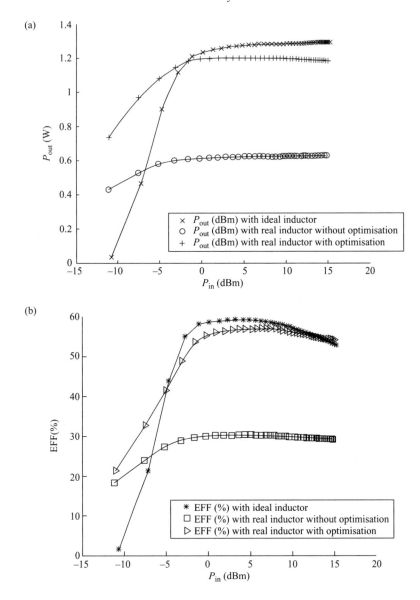

Figure 9.23   Simulation results with ideal inductors, and parasitic-laden inductors before and after optimisation

power is 1.2 W and the drain efficiency is 58 per cent for an input power level of 0 dBm. Replacing the ideal inductors with their parasitic-laden spiral and bond-wire counterparts, but before optimisation, the output power drops to 0.6 W and the drain efficiency decreases to only 30 per cent with 0 dBm of input power.

Table 9.3 *Summary of output power, drain efficiency and gain before and after optimisation of the three-stage RF power amplifier*

Specification	Ideal inductors	Real inductors	
		without optimisation	with optimisation
Input power	0 dBm	0 dBm	0 dBm
Output power	1.2W	0.6W	1.2W
Gain (@ $P_{in} = 5$ dBm)	30 dB	27 dB	30 dB
Drain efficiency	58%	30%	55%

Clearly, the parasitics of the passive components significantly degrade the performance of the power amplifier by effectively de-tuning it. Also, Figure 9.23 shows the simulation results with parasitic-laden inductors after parasitic-aware optimisation. As illustrated, parasitic-aware RF synthesis is essential in the design of efficient high-performance RF circuits that are robust to process, temperature and voltage variations. As summarized in Table 9.3, the use of CAD optimisation significantly improves circuit performance; the final design including passive parasitics meets the original design specifications.

9.4.3 CMOS RF distributed amplifiers

The four-stage distributed amplifier shown in Figure 9.24 employs two artificial LC delay lines. One, called the *gate line*, applies delayed versions of the input RF signal sequentially to the four gate terminals [11, 12]. The other, *the drain line*, adds the drain signal currents constructively in the load resistor. Cascode transistors are used to reduce the Miller effect by imposing low impedance at the drains of the amplifying devices. Cascoding also reduces capacitive coupling between the artificial transmission lines and increases gain flatness, reverse isolation, stability and input and output impedance matching accuracy. The high output impedance of the cascode configuration also reduces loss associated with amplifier loading on the drain line.

An LC transmission line exhibits an intrinsic mismatch at each termination point due to image impedance variations with frequency. Hence, *m*-derived half-sections are inserted to improve the impedance matches to the delay lines.

In order to increase gain flatness and decrease gain peaking near the cut-off frequency, a staggering technique is frequently used in distributed amplifiers. Staggering basically means adopting slightly different cut-off frequencies for each delay line in order to increase the overall linearity of phase response. Specifically, designing the gate line to have a slightly higher cut-off improves gain flatness. Assuming matched impedances at the input and output ports of the distributed amplifier, the cut-off

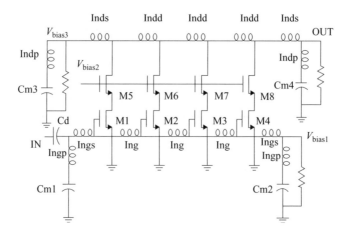

Figure 9.24 Four-stage CMOS distributed amplifier with artificial LC gate and drain delay lines

frequency, characteristic impedance and low-frequency gain are:

$$f_c = \frac{1}{\pi \sqrt{LC}} \qquad (9.13)$$

$$Z_0 = \sqrt{\frac{L}{C}} \qquad (9.14)$$

$$S_{21} \cong \frac{n}{2} g_m Z_0 \qquad (9.15)$$

where g_m is the small-signal transconductance of the identical active devices. The n term in (9.15) indicates the unique property of gain addition in the distributed amplifier.

The technology used is a 0.35 μm CMOS process. Key amplifier specifications are a constant gain greater than 6 dB over a bandwidth greater than 6 GHz with linear phase over the full bandwidth. The input and output ports are matched to 50 Ω, and the distributed amplifier operates from a single 2.5 V power supply.

In the distributed amplifier, 11 design parameters are optimised with a cost function given by

$$\text{Cost} = \frac{\text{abs(gain} - 2)}{2} \cdot \text{weight\_gain} + \frac{1/(\text{DIFF\_gain}) - (1/0.6)}{1/0.6} \cdot \text{weight\_DIFF}. \qquad (9.16)$$

where weight_gain and weight_DIFF are weight functions of gain and DIFF terms.

According to equation (9.16), the gain needs to be greater than 2 (6 dB), and the difference (DIFF_gain) between maxgain and mingain needs to be less than 0.6. In this simulation, weight_gain and weight_DIFF are 50.

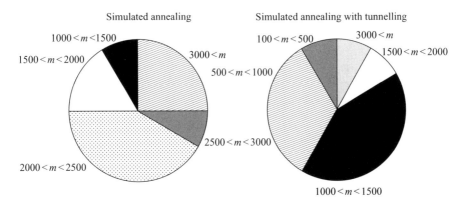

Figure 9.25   The number of iterations required to find the optimum solution for simulated annealing without and with tunnelling (12 synthesis runs) for the distributed amplifier

9.4.4   Distributed amplifier optimisation results

Figure 9.25 shows the comparison between simulated annealing and adaptive simulated annealing with tunnelling algorithms. Because there are only 11 design parameters to be optimised in this distributed amplifier, both methods find the solution within 5,000 iterations. Yet, as shown previously in the power amplifier case, adaptive simulated annealing finds the solution with fewer iterations than simulated annealing. As shown in Figure 9.26, the local optimisation and tunnelling techniques save a substantial number of iterations.

Table 9.4 shows the design parameter values identified in Figure 9.24 before and after optimisation. As shown, many inductors and capacitors become smaller because the design now *takes advantage* of their parasitics as part of the design process.

As was the case with the power amplifier, Figure 9.27 and Table 9.5 also show an intolerable degradation in distributed amplifier performance when the ideal inductors are replaced by their parasitic-laden counterparts: the bandwidth decreases to about 2 GHz, the gain drops to 3 dB, and the flatness of the gain characteristic is lost. In this example, parasitics adversely affect the final design because they were not considered early in the design phase. Conversely, the parasitic-aware design methodology considers parasitics as an integral part of the synthesis process. Efficient CAD synthesis is the key enabling technology for *parasitic-aware methodology*.

9.5   Conclusions

The parasitic-aware design and optimisation of RF circuits has been discussed. Parasitic-aware CAD synthesis is essential in the design of CMOS RF

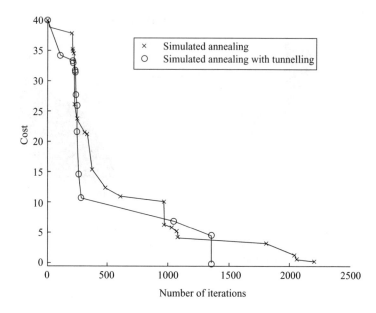

Figure 9.26   Cost functions versus the number of iterations in conventional and adaptive simulated annealing with tunnelling

Table 9.4   Summary of the 11 design parameters before optimisation and after optimisation with a distributed amplifier (see Figure 9.24)

Value	Ideal	After optimisation
Ingp	0.743 nH	0.23 nH
Indp	1.06 nH	0.73 nH
Ings	1.114 nH	1.273 nH
Ing	1.39 nH	1.59 nH
Inds	1.59 nH	1.77 nH
Indd	1.99 nH	2.68 nH
Cd	600 fF	186 fF
Cm1	167 fF	144 fF
Cm2	167 fF	431 fF
Cm3	239 fF	713 fF
Cm4	239 fF	0 fF

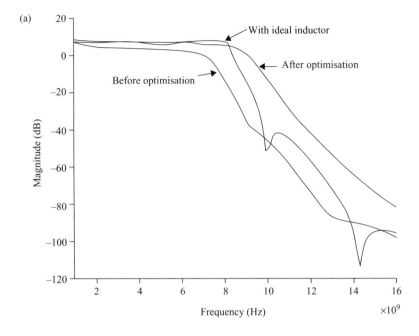

(a)

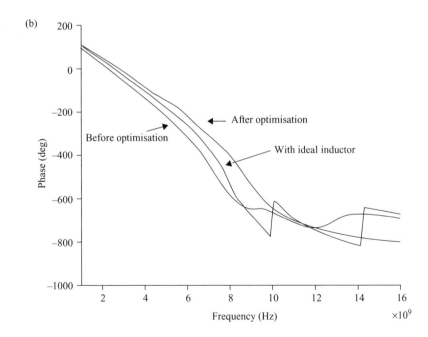

(b)

Figure 9.27   *(a) Forward gain* $(S_{21}(dB))$ *magnitude and (b) forward gain phase results*

Table 9.5　Summary of bandwidth, gain and gain flatness before optimisation and after optimisation with distributed amplifier

Specification	Ideal inductors	Real inductors	
		without optimisation	with optimisation
Bandwidth	8.4 GHz	7.0 GHz	9.1 GHz
Average gain	7.55 dB	3.54 dB	6.37 dB
Flatness	0.5 dB	3.2 dB	0.96 dB

communications circuits due to the parasitic effects that must be considered early on as part of the design.

An adaptive simulated annealing method with tunnelling and a post-optimisation PVT analysis technique have been proposed. Examples show that adaptive simulated annealing accelerates the optimisation as well as escaping from local minima more easily than with simulated annealing. Also, the post-optimisation PVT method was able to consider PVT variations in circuit optimisation with a minimal increase in iteration count.

9.6　References

1　GREENHOUSE, H. M.: 'Design of planar rectangular microelectronic inductors', *IEEE Transactions on Parts, Hybrids, and Packaging*, **10** (2), June 1974, pp. 101–09

2　CRANINCKX, J. and STEYAERT, M. S. J.: 'A 1.8-GHz CMOS low-phase-noise voltage-controlled oscillator with prescaler', *IEEE Journal of Solid-State Circuits*, Dec. 1995, **30**, pp. 1474–82

3　LEE, Y.-G., YUN, S.-K. and LEE, H.-Y.: 'Novel high-Q bond wire inductor for MMIC'. IEEE International Electron Devices Meeting, 1998, pp. 19.7.1–19.7.4

4　RUTENBAR, R. A.: 'Simulated annealing algorithms: An overview', *IEEE Circuits and Devices Magazine*, Jan. 1989, **5** (1), pp. 19–26

5　GUPTA, R. and ALLSTOT, D. J.: 'Parasitic-aware design and optimisation of CMOS RF integrated circuits', *IEEE RFIC Symposium*, June 1998, pp. 325–28

6　GUPTA, R. and ALLSTOT, D. J.: 'Parasitic-aware design and optimisation of CMOS RF integrated circuits', *IEEE International Microwave Symposium*, June 1998, pp. 1847–50

7　GALLEGO, R. A., ROMERO, R. and MONTICELLI, A. J.: 'Tabu search algorithm for network synthesis', *IEEE Transactions on Power Systems*, May 2000, **15**, pp. 490–95

8 CHOI, K., ALLSTOT, D. and KIAEI, S.: 'Parasitic-aware synthesis of RF CMOS switching power amplifiers'. International Symposium on Circuits and Systems, May 2002, pp. 269–72

9 SOKAL, N. O. and SOKAL, A. D.: 'Class-E.: A new class of high-efficiency tuned single-ended switching power amplifiers', *IEEE Journal of Solid-State Circuits*, June 1975, **10** (3), pp. 168–76

10 RAAB, F. H.: 'Idealized operation of the class-E tuned power amplifier', *IEEE Transactions on Circuits and Systems*, December 1977, **24** (12), pp. 725–35

11 SARMA, D. G.: 'On distributed amplification'. Proceedings of Institution of Electrical Engineers, 1954, **102B**, pp. 689–97

12 GINZTON, E. L. et al.: 'Distributed amplification'. Proceedings of IRE, Aug. 1948, **36**, pp. 956–69

Acknowledgements

We are pleased to acknowledge support for the research described herein from the following sources: DARPA NeoCad Program under grant N66001-01-8919; Semiconductor Research Corporation grants 2001-HJ-926 and 2003-TJ-1093; National Science Foundation contracts CCR-0086032, CCR-01200255 and MRI-0116281, and grants from the National Science Foundation Center for the Design of Analog/Digital Integrated Circuits (CDADIC), Texas Instruments, National Semiconductor, and Intel.

Chapter 10

Testing of RF, analogue and mixed-signal circuits for communications: an embedded approach

Mohamed M. Hafed and Gordon W. Roberts

10.1 Introduction

The importance of testing of semiconductor devices in general comes about because of the decidedly imperfect nature of the manufacturing process and its associated tolerances. Given the fine pitch of modern semiconductor devices, even a few particles of dust or debris that fall on a device during fabrication can permanently damage it. As it is impossible to completely eliminate such defects or contaminants from the wafer fabrication and manipulation process, a certain percentage of parts coming out of a fabrication run will always be defective, and a production test phase is required to screen such parts. Apart from debris particles or wafer defects, the variability of the fabrication process itself can also significantly degrade the 'quality' of a fabricated device. As device dimensions approach the atomic size [1], even single-atom disturbances can result in significant alteration to the electrical parameters of a device. Low-speed digital design techniques in CMOS are forgiving of electrical parameter variations [2]; however, high-speed digital as well as analogue and RF techniques are sensitive to such variations. Thus, for digital circuits, the problem of production testing is not that of performance measurement or of characterising electrical parameter variation. Instead, it is that of finding structural alterations such as short-circuits or open-circuits due to physical contaminants. When 'analogue' performance metrics such as the maximum operating frequency of a microprocessor are required, the test problem changes significantly since a defect-free device can still fail a performance metric. In other words, there exists an uncertainty in the performance of fabricated devices even in the absence of wafer defects or unwanted debris particles, and some direct performance-based test techniques are required [3].

The other main requirement in any production test strategy is speed of test execution and device separation. It is this speed requirement that makes production testing of analogue circuits so challenging. The reason is that analogue circuits are defined by a large set of specifications. As will be seen shortly, attempting to verify all specifications in the production test phase is too prohibitive, so careful planning of which specifications to test is important. In this chapter we describe both techniques for such test selection and optimisation as well as hardware techniques for test-signal delivery. We also describe a recent trend in test integration which promises to reduce many of the burdens of analogue testing. Specifically, we describe the use of embedded mixed-signal test cores, which are integrated circuit 'macros' that emulate the functions of fully-fledged automatic test equipment. These embedded test cores are designed to perform DC curve tracing, oscilloscope, timing, and frequency domain measurements using compact and mostly digital integrated electronics. Such scaling of measurement instruments is envisioned to be a natural consequence of semiconductor scaling trends [1] and of performance, cost, and signal handling pressures that are starting to render external test equipment centric techniques more and more impractical.

The scope of the techniques in this chapter is all the analogue and mixed-signal circuitry in a radio IC. This includes the baseband and frontend circuits as well as mixed-mode components such as data converters and clock circuitry.

10.2   Some analogue performance metrics and test fundamentals

We introduce some of the analogue performance criteria for wireless communications devices in the context of the simplistic sample signals shown in Figure 10.1. Since many of the concepts presented in this section have been covered earlier, a qualitative description of this figure should suffice to review some of these concepts and describe their influence on test. For example, Figure 10.1a suggests that a radio receiver is expected to decipher a sometimes very weak user channel in the presence

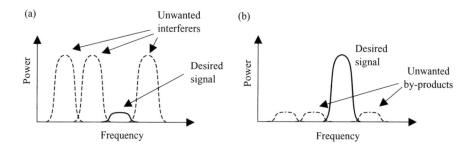

Figure 10.1   *Sample signal spectra at (a) the input of a receiver and (b) the output of a transmitter. Linearity, frequency selectivity, noise performance and other performance requirements can easily be described through direct observation of such spectra*

of large adjacent channels. The result is stringent linearity requirements not only at the frontend circuits but also downstream through the intermediate frequency circuits and/or the digitising circuitry. On the other hand (Figure 10.1b), a transmitter is ideally required to generate only the user signal without any spurious by-products due to circuit non-linearity. The goal of testing in both of these cases is to determine whether such signal discrimination is actually performed by the device under test (DUT). Figure 10.1 also suggests that frequency selectivity is an equally important function in radio transceiver ICs which are expected to selectively process the user channel while suppressing some strong interfering channels. Depending on the architecture used, frequency selectivity is performed at different places (and different frequency bands) within a single transceiver device, the ultimate goal being to separate the user channel from all other channels. Some other performance requirements for wireless communications devices, such as noise performance, can similarly be inferred from Figure 10.1.

The above description of some of the behavioural requirements for wireless communications devices translate into corresponding circuit-specific requirements on linearity, spectral purity, noise, etc. Some of the metrics for these parameters are described in more detail here.

10.2.1 Linearity

When a signal passes through a non-linear system, it can get adversely modified by the input–output characteristic of the system. Assuming a memoryless system, a convenient method for representing the behaviour of the output signal is to expand it into a power series representation:

$$v_o = a_0 + a_1 v_{in} + a_2 v_{in}^2 + a_3 v_{in}^3 + \cdots . \tag{10.1}$$

Linear and non-linear circuit behaviour is available directly from a frequency domain representation of the output signal of a circuit when excited by a sinusoidal signal. Specifically, consider a circuit that is described by the relation (10.1). If this circuit is excited by a sinusoid as

$$v_{in} = A \cos(\omega t) \tag{10.2}$$

then its output can be expressed as

$$v_o = a_0 + a_1 A \cos(\omega t) + a_2 A^2 \cos^2(\omega t) + a_3 A^3 \cos^3(\omega t) \tag{10.3}$$

where only the first three terms of (10.1) are retained. Rearranging (10.3), the output can be written as

$$v_o = a_0 + \tfrac{1}{2} a_2 A^2 + \left(a_1 A + \tfrac{3}{4} a_3 A^3 \right) \cos(\omega t)$$
$$+ \tfrac{1}{2} a_2 A^2 \cos(2\omega t) + \tfrac{1}{4} a_3 A^3 \cos(3\omega t). \tag{10.4}$$

Equation (10.4) gives insights into many of the parameters that are usually sought in linear circuit specification. First, a single sine wave at the input of a non-linear system results in several waves at the output, the fundamental signal at the same frequency as the input and several harmonically related sine waves. The number and power of the extra harmonics is referred to as harmonic distortion, and it is one of the most important linearity tests for analogue circuits. Harmonic distortion, for example, is a key parameter in evaluating the baseband components such as D/A or A/D converters in a transceiver. Similarly, observation of (10.4) gives insights into how non-linearity modifies the amplitude of the original input signal. Apart from the desired gain introduced by the circuit, its non-linear nature introduces an additional term at the same frequency as that of the input signal. This phenomenon is known as gain compression, as this additional term usually acts to reduce the amplitude of the fundamental output signal. A common metric for gain compression is the 1 dB compression point, which is the input amplitude level that causes the circuit gain to decrease by 1 dB. From a test perspective, measuring this parameter requires delivering a spectrally pure sinusoidal signal of *known* amplitude to the linear component under test and extracting the output signal power. Another parameter that is related to gain compression is desensitisation, and it becomes important when more than just a sine wave is being processed. Specifically, when both a weak and a strong signal are input to the circuit, gain compression due to the strong signal acts to reduce the circuit's gain and, hence, the ability to amplify the weak signal.

Of special interest in narrowband systems is an intermodulation test, which is a yet more realistic measure of linearity. When more than just a single sine wave is applied to (10.1), it can be shown that the output exhibits sine wave components that are not harmonics of the input frequencies. Such components, which are called intermodulation products, can fall very close to the input signal frequencies and can, thus, reveal information about the non-linearity of the system in a narrow frequency band. Referring to Figure 10.1, intermodulation is actually the mechanism by which interfering channels affect an adjacent user channel. Again, from a test point of view, *simultaneous* sine wave generation with proper amplitude control is required. A typical parameter for measuring intermodulation is called the third-order intercept point. To obtain this point, a two-tone signal is applied to the circuit under test and the output fundamental two-tone signal as well as the third-order intermodulation products are measured. The third-order intercept point is defined as the input two-tone amplitude level for which the power in the intermodulation products exceeds the power in the fundamental output of the system.

Other linearity tests that are also important for analogue circuits include DC characteristics and common-mode and offset characteristics. Some of these are described briefly later in the text.

10.2.2   *Frequency response*

Different communication standards pose different requirements on the filtering properties of a radio system [4]. Similarly, different implementation choices also affect the filtering requirements in a radio transceiver. For example, Reference 5 describes recent trends in data converter design for radio applications that pose a wide range of analogue filter requirements. In general, both narrowband bandpass filters as well as lowpass anti-image or anti-alias filters are required.

The most important test for the frequency selective components of the radio IC is the magnitude and phase frequency response. Traditionally, a single sine wave with a swept frequency is applied to the filter under test, and the filter output magnitude and phase are recorded. With the recent application of DSP-based techniques in the test and measurement world, a much faster method for obtaining the same frequency response information is to excite the filter with several sine waves simultaneously and to digitise the output of the filter and process the magnitude and phase of each of the output sine waves. Several examples of this multitone test for analogue filters will be described in this chapter. With this approach, however, the choice of the input sine wave frequencies is important. Because of intermodulation due to the non-linearity of the filter, the filter output will contain not only the input component frequencies but also many other components due to the intermodulation terms. In order to avoid errors in the gain and phase measurement at the desired test signal frequencies, the latter have to be chosen to avoid the coexistence of intermodulation terms at the same frequencies as the test sine waves. Fortunately, several software methods for component frequency selection exist in the literature as described in Reference 6. The advantage of such an approach to filter response measurement is that other useful filter performance metrics, such as group delay, can easily be obtained through the application of DSP algorithms on the digitised filter output signal.

10.2.3   Noise

Because wireless transceiver circuits are required to operate on extremely low signal levels, the effects of noise due to the circuits themselves and due to external interference sources are given careful consideration in the design of wireless transceivers [4]. The treatment of noise is beyond the scope of this chapter and will not be included here. Instead, we introduce the concepts of the signal-to-noise ratio (SNR) and signal-to-noise and distortion ratio (SNDR). As can be expected from these terms, both are measures of the ratio of a certain desired signal's power to the power in noise (or noise plus distortion) that accompanies this signal. Such ratios are important at different locations within a circuit (e.g. input and output) or within a group of circuits (e.g. in a cascade of circuits), and they are especially challenging for test and characterisation. For example, in measuring the SNR at the output of a linear amplifier, care has to be taken so that the measurement instrument itself does not introduce excessive noise thus corrupting the estimate of the circuit's own noise generation properties. In modern test instruments, DSP methods are used to estimate noise power spectra in digitised circuit response signals. Such methods will be described briefly in subsequent sections.

Having the ability to compare signal levels to noise levels provides other useful information about analogue circuits and their ability to deal with random phenomena. For example, the ratio of the SNR at the input of a circuit to the SNR at its output is referred to as the noise figure, and it is a widely used metric in RF design. Similarly, by measuring the SNR at different power levels, the dynamic range of a circuit (or spurious-free dynamic range, SFDR) can be obtained as a measure of the range of signals that can be processed by a device [6].

10.2.4   Other performance metrics

The above sections introduced some of the important specification metrics for analogue and mixed-signal circuits. However, the complete specification set for such devices is more diverse [6]. For example, other important parameters include DC parametric performance, settling time, overshoot, cross-talk, jitter and substrate noise. Some of these parameters are increasingly important in a SOC environment in which large digital blocks are integrated alongside sensitive analogue blocks. These specifications also pose more diverse test requirements, and related signal delivery and extraction mechanisms, which are at the heart of the topics described in this chapter.

10.3   Translating design specifications into production test programmes

Much of the recent mixed-signal test research falls under the category of reducing test time. The reason is that specification testing, although foolproof, can be extremely time-consuming. Section 10.2 described some of the analogue specifications for mixed-signal circuits. A quick look at the data sheet of a commercial mixed-signal device would reveal that these as well as other specifications also span a wide range of input and environmental (such as power supply and operating temperature) conditions. Testing all of these conditions for all specifications during production would be slow. The question then is whether a simplified set of signals or input conditions can be used to excite the DUT and uncover most of the failures that could occur during fabrication.

Traditionally, this question has been answered through an empirical approach in which a reasonably large number of fabricated devices are subjected to full specification tests. The results of this procedure are then used to compile a short list of tests that fail most of the defective parts. Over the rest of the product life for the part, only the tests in this trimmed list are applied to the DUT and used to screen the faulty devices, and redundant tests are simply omitted. Moreover, it is also conceivable that there be significant correlation between tests [7] and that more than one test fails the same faulty device. Recognising this, early work in mixed-signal circuit testing consisted of methods for not only reducing the test list in the manner just described but for also choosing an order in which tests are applied. As such, the test that is most likely to fail is applied first, followed by less likely ones. Various methods for ordering and trimming the tests include those in References 8–10. Beyond the above methods, a significant portion of the published literature on mixed-signal testing attempts to perform such test list trimming without actually exhaustively testing a subset of the fabricated devices. These approaches can be described within the scope of the flow diagram of Figure 10.2, although depending on the method, one or more of the nodes

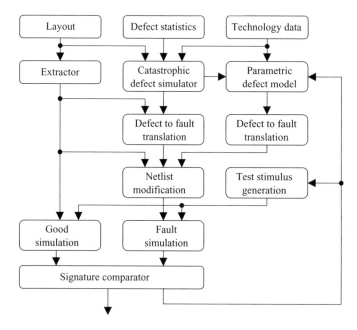

*Figure 10.2   Generic flow of a simulation-based defect-oriented test methodology.
Several variations exist. Problems with this approach include a general
disagreement on the implementation of each of the steps in this methodology as well as its applicability to predicting post-manufacturing
effects*

in this diagram might not be needed. Assume for the time being that we know all
defects that can affect the DUT. In these methods, the layout of the circuit under
test is modified according to the effects of these defects, the new extracted circuit is
simulated, and the response of this new 'faulty' circuit is compared to the response
of the original one. Depending on the input conditions and the resulting response
comparison, some test signals can be found to be 'useful' in detecting manufacturing defects, while others are deemed not useful [7]. As can be seen, the result of
the above procedure is a *short* set of test stimuli (DC, AC, or transient) that need
to be applied to a DUT during the actual test phase and a second set of expected
'signatures' or circuit responses that can distinguish between a good part and a bad
part. The size of the set is usually kept to a minimum in these techniques. For example, in Reference 11, only a few DC levels out of thousands are used to test an A/D
converter.

While conceptually appealing, the above procedure for test selection is marred
with hurdles that have hindered its progress over the past. These include the difficulty
in defining what a defect or fault is for analogue components [12] and modelling inac-
curacy associated with simulating the virtual defective devices [13]. Several attempts
have been made to address these difficulties with varying degrees of success. The

interested reader is referred to References 9, 10, 14, 15 for different approaches to modelling defects and their effects on analogue circuits and to References 16–18 for some approaches to simulating the effects of these defects. Finally, the test stimulus generation node in Figure 10.2 has also been the subject of extensive research. Two representative approaches to test pattern generation are included in References 19 and 20.

10.4   The (diminishing) role of test equipment

Regardless of how test list selection is performed or whether only defects should be targeted as opposed to specification testing, the ability to deliver signals reliably to the DUT has never been under such scrutiny as it is right now. The physical limits of interconnects between instrument and DUT and the disparate nature of DUT and instrument result in serious difficulties in modern-day test arrangements, whether on validation benches or on production floors. This section describes the architecture and operation of test equipment as well as some of the challenges they face in the context of high-frequency analogue testing such as in the case of wireless communications devices.

Modern test instruments almost invariably rely on powerful DSP techniques in order to facilitate automation [21] and to enhance measurement accuracy and repeatability. As was mentioned in Section 10.2, using DSP techniques, a DUT response signal is measured by first digitising it using a highly accurate analogue-to-digital (A/D) converter. Subsequently, microprocessor based computations are performed in order to analyse the digitised signals. For example, in a spectrum analyser, a block diagram of which is illustrated in Figure 10.3a, hardware or software implementations of a Fast Fourier Transform (FFT) are utilised before displaying the results to a user monitor. As one proceeds to higher test signal frequencies, an extension to the diagram of Figure 10.3a is required as shown in Figure 10.3b. In such an extension, the high-frequency signal is first transported to a lower frequency by passing it through a mixer that is driven by a spectrally pure local oscillator (LO). The lower frequency

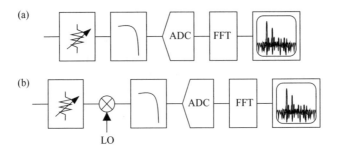

Figure 10.3   *(a) A DSP-based spectrum analyser. (b) Similar spectrum analyser capable of characterising higher-frequency signals by shifting them down to the frequency range of the ADC*

is chosen to fall within the range that the A/D converter can tolerate without compromising accuracy. Moving a bit higher in frequency, further modifications to the basic block diagram of Figure 10.3a are necessary. Specifically, at sufficiently high frequencies, wave propagation phenomena across interconnections between instrument and DUT have to be considered, and since perfectly matching input and output impedances across all frequencies is generally impossible, it becomes imperative to evaluate signal losses due to reflections at the pertinent interfaces. To perform such tests, directional couplers are incorporated in network analysers in order to separate incident and reflected waves at a given port [21]. Network analysers are currently indispensable in the test and characterisation of large-volume radio-frequency (RF) ICs.

While providing extremely high reliability and programmability, the above DSP-based instruments are still relatively cumbersome to automate because of size and cost issues. They have thus remained on the validation and characterisation bench. Historically, the path to the interconnection and synchronisation of a multitude of equipment for production floors evolved into building automatic test equipment (ATE). Specifically, many of the underlying test-specific components (and computations) of bench instruments are replaced in ATE by centralized processing, synchronisation, and control. A single programming language and syntax are used for seamlessly operating the various instruments. The actual 'measurement' portions of the hardware described in Figure 10.3 are also built into the ATE using off-the-shelf components. Figure 10.4 shows a functional diagram of a typical mixed-signal automatic tester [21]. As can be seen, the other main difference to note from bench equipment is in signal handling. Due to the need for automation and general-purpose functionality, a significant effort goes into providing efficient signal handling mechanisms. A more detailed description of the generic architecture of ATE and its limitations as well as a description of signal handling are both described in the following subsections.

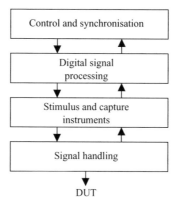

Figure 10.4   ATE integration of DSP-based measurement instruments. Central processing and control are used, and an automatic signal handling mechanism is incorporated

10.4.1   *Generic architecture of automatic test equipment*

A more detailed view of a mixed-signal automatic tester is shown in Figure 10.5. It includes various sources and meters, an extensive relay matrix, clock and synchronisation sources, timing measurement devices, and a digital subsystem for digital test vector formatting, comparison and manipulation. In order to remain a general-purpose instrument, a mixed-signal tester relies on a few components, such as arbitrary waveform generators and digitisers, and flexible DSP techniques that can cover many tests and span different DUT specifications. In general, the analogue components in the above architecture are the critical ones since they potentially operate on low-level signals and since their function (whether sourcing or capturing) is altered by the physical interaction with whatever is connected to their interface terminals.

When more analogue channels need to be tested on the same platform, more of the specialised analogue instruments (arbitrary waveform generators and digitisers) are incorporated within the test system. This is one of the cost drivers for RF ATE, since the analogue instruments themselves are costly. Moreover, the new instruments have to be synchronised with the digital components of the tester, powered up and cooled adequately, and seamlessly incorporated into the tester's software environment. Such difficulties have meant that a limited number of instruments could be incorporated into a tester platform, which in turn means that a limited amount of test parallelism can be exploited to maximise test throughput during production. Typically, only two or four analogue channels are tested simultaneously on a single ATE during production.

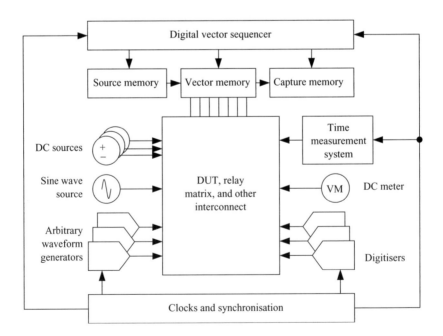

Figure 10.5   Block diagram of a generic mixed-signal automatic tester [6]

10.4.2   Signal delivery

The portion of the ATE generally responsible for signal delivery is the test head, and it is where components like directional couplers, pin electronics and DUT-related power supplies are found. Also on the test head is a device interface board (DIB), which is a multilayer printed circuit board (PCB) responsible for implementing the final connection between instruments internal to the ATE and the DUT. The greatest difficulty in the initial phases of production testing is usually in ensuring that the DIB behaves in a manner that closely mimics conditions used during device characterisation [6]. The reason is that the test engineer, not the ATE vendor, is usually responsible for the design of this board. Moreover, instrument performance specifications provided by the ATE documentation do not generally assume signal delivery through the DIB. It is the responsibility of the DIB designer and/or test engineer to ensure that the measurement specifications are met even as the signals propagate through the DIB. This is a task that is easier said than done. The primary reason is that signal integrity analysis (with full field solvers) is generally difficult. Instead, DIB design relies on experience, prototyping, and rules of thumb. Some attempts have been made to provide simulation tools for the DIB or the complete signal paths from the pin electronics to the DUT. A good example of this is the approach described in Reference 22. In this approach, the problem is simplified by tackling each of the connectors, sockets, cables, contactors and PCB traces separately. Moreover, a combination of measurement techniques such as time-domain reflectometry and simulation techniques using field solvers is employed.

10.5   Embedded mixed-signal test cores

In light of the discussion in Section 10.4, we note that many of the challenges facing the continued usefulness of automatic test equipment are related to limited integration levels. Specifically, it was shown that limited parallelism is expected to exist on a test platform unless major integration efforts are undertaken by the ATE vendor. Similarly, it was mentioned that traditional test setups place a DUT in an environment that is very different from its mission environment. For high-frequency parts that are not expected to drive long interconnects in the field, extracting realistic device performance data in such an environment becomes extremely challenging. In order to deal with some of these challenges, design-for-test approaches that enhance the testability of mixed-signal devices have been described in the literature. However, as described in Reference 23 and elsewhere, such approaches are either tailored to a specific circuit and constitute clever modifications to such a circuit, or they require that some components such as data converters already exist on the device. Instead, the goal of this section is to describe how the general-purpose functionality of external DSP-enabled test equipment can be emulated on-chip without relying on the existence of certain components inside the DUT. Thus, the circuits described here can be thought of as miniature versions of some of the critical analogue instruments on ATE. Such miniaturisation promises to provide a significant improvement in the number of analogue test channels available for simultaneous operation. Thus, for sufficiently complex devices, many blocks within the devices can be tested simultaneously. Alternatively, multisite testing in which several devices are tested simultaneously can be facilitated since it now does not rely on the limited number of instruments available on the tester. Another goal of this section and the next ones is to describe how providing such flexible test capability on-chip also avoids interconnect-related difficulties and provides much improved correlation between simulation, design verification, and production testing. Improved correlation comes about, of course, because the same measurement instruments and input conditions are used in simulation, design validation and production testing.

Functionally, our proposed integrated test core looks like a fully-fledged generic DSP-based test system (Figure 10.6). It consists of a band-limited arbitrary waveform generator and a periodic waveform digitiser both synchronised using a single clock source. A more detailed view of the components comprising our test system is shown in Figure 10.7, although the clocking system is omitted for clarity. The system consists

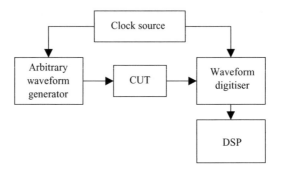

Figure 10.6   Block diagram of proposed on-chip test system

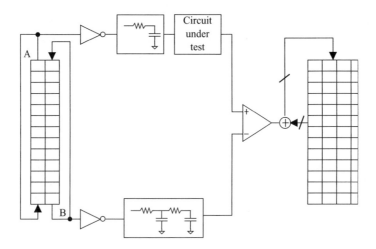

Figure 10.7   A more detailed view of the integrated test system architecture

of two one-bit memory files (referred to as Memory A and Memory B), some simple analogue structures, and a multibit output memory for output data storage. A key feature of this architecture is its simplicity and modularity: an almost all memory (all digital) implementation is used, which allows for faster design times, robustness to process variation, and scalability. The left part of the figure consists of two one-bit memory-based analogue signal generators, one for AC stimulus generation and one for DC voltage generation. The latter is combined with the analogue comparator to perform multibit-resolution digitisation. The following subsections describe the encoding methods for the signal generators and our signal capture method.

10.5.1 Signal generation (circular memory A)

We synthesise analogue test signals using a short repetitious sequence of digital bits that are chosen to approximate the output of a one-bit $\Sigma\Delta$ modulator when driven by a periodic signal. The concept behind this generation approach was described in References 24 and 25, and it consists of simulating a software model of a high-order noise-shaping modulator and capturing a finite duration segment of its output. Periodically repeating this segment approximates the usually aperiodic infinite-duration output of the $\Sigma\Delta$ modulator. However, as described in Reference 25, in order for the periodic approximation to achieve a high fidelity, the frequency of the input signal to the software $\Sigma\Delta$ modulator has to be harmonically related to the chosen fundamental frequency of the bit sequence. If N is the length of the approximate $\Sigma\Delta$ sequence and F_S is the sampling rate, then the input to the modulator has to be an integer multiple of F_S/N. In fact, the forced periodicity of the approximate $\Sigma\Delta$ output also means that the encoded signal contains only multiples of this same fundamental frequency.

As can be seen, the frequency resolution of this approach, which is defined as the minimum frequency deviation between two simultaneously generated sine waves, is thus F_S/N. Also, the clock-derived frequency synthesis property of this memory-based generator is an important feature that makes it favourable in our application as it guarantees sample coherence with our on-chip circuit-response digitiser. Coherent sampling enables the use of a small number of samples in a DSP-based measurement environment [26].

Since our periodic bit patterns approximate the output of a $\Sigma\Delta$ modulator, general properties of $\Sigma\Delta$ modulation determine the quality of the resulting analogue signals. For example, just like $\Sigma\Delta$ modulators used in data conversion applications, increasing modulator order has the promise of improving in-band SNR [27]. Other factors that determine the performance of $\Sigma\Delta$ modulators include modulator topology and the oversampling ratio (OSR). As for the periodic $\Sigma\Delta$ approximations considered here, these parameters as well as other tools such as the choice of bit sequence length, arbitrary choice of centre frequency, and unconventional stability properties [25] are *all* available for obtaining the best bit streams for the application of interest. Indeed, the possibility of modifying encoded analogue signal properties through a manipulation of these tools results in tremendous versatility using this approach.

As an example of the quality of signals encoded using this method, Figure 10.8 illustrates the power spectral density (PSD) of a third-order noise-shaped periodic

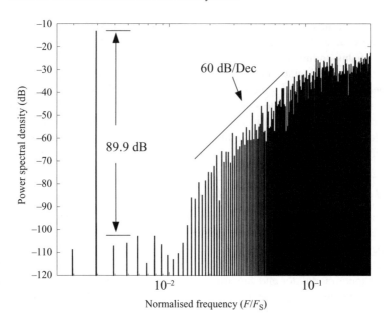

Figure 10.8   Frequency spectrum of a periodically repeating $\Sigma\Delta$ modulated sequence encoding a single sine wave. The sequence length in this example was 1024

$\Sigma\Delta$ stream. As can be seen, noise due to the one-bit quantisation operation on the input sine wave is shaped out of band in much the same way as would be the case in a $\Sigma\Delta$ modulator. The tonal behaviour of the noise is simply an outcome of the periodic nature of the bit stream. In general, the signal generation operation is completed by filtering out these noise harmonics and reconstructing the original sine wave. A description of the trade-offs involved in the design of the filter as well as applications in which such a filter may or may not be required is included in [28] and is omitted here in the interest of brevity. In a nutshell, the ability to modify the spectral properties of the periodic $\Sigma\Delta$ streams results in fairly relaxed requirements on the analogue reconstruction filter. Rather than designing the filter to match some encoding scheme, we instead shape the spectrum of the programmable streams to match the roll-off behaviour of the implemented filter within the desired test system bandwidth. Alternatively, the device being measured might itself contain a filtering operation as is the case with many of the building blocks in a wireless transceiver circuit, and an additional filter is not required at all in such a scenario. We are of the view that separating the filter design in this case from the synthesis of the overall test system allows for a more flexible and modular solution. Coming back to Figure 10.8, we specifically show a low-order example in this figure since low-order noise shaping relaxes the requirements on the order of any analogue filter that reconstructs the encoded in-band signal. Even with such a low order, high spectral purity is achieved using only a 1024 bit long memory. Examples of high-frequency analogue signal generation through the use of bandpass modulation in encoding the periodic $\Sigma\Delta$ streams [25] are illustrated in Section 10.6.

10.5.2 Signal digitisation (circular memory B and comparator)

Signal digitisation is performed using the combination of circular memory B, its associated passive RC filter, and the voltage comparator. The combination of these blocks in the manner shown in Figure 10.7 enables us to achieve multibit analogue waveform digitisation using compact and simple hardware. This is done by exploiting the periodicity of the analogue signal under test (which is enforced in our architecture) and making multiple comparison passes over progressive periods of this signal. Referring to Figure 10.9, the complete digitisation process proceeds as follows [29]. First, the reference input to the comparator is set to a constant voltage to which all samples of a unit test period (UTP) of the analogue waveform are compared. This voltage of course represents one of the quantisation levels of the overall A/D conversion. Once all comparisons to this level are made (stored in memory), the voltage at the reference input to the comparator is incremented to the next level, and the process repeats. As an alternative, we can use a binary search algorithm over progressive runs of the UTP [30], but the principle of successively approximating the analogue waveform over multiple UTP runs remains the same.

The way the reference input to the comparator is defined in our system is through the combination of memory B and its associated averaging filter. Specifically, accurate DC reference levels are encoded in memory B, again using a periodic sequence

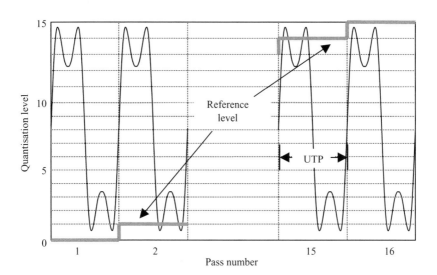

Figure 10.9 *Exploiting the periodicity of the input signal to perform multiple digitisation passes. UTP = Unit Test Period*

of digital $\Sigma\Delta$ modulated bits whose average equals the desired DC level [28, 31]. By encoding an average signal in a periodic rectangular waveform, extremely high linearity is achieved. This is an important requirement in our application since the linearity of the overall on-chip digitiser is determined by the linearity of the reference inputs to the comparator. In other words, the combination of the comparator and a variable reference generator is similar in principle to a successive-approximation A/D converter, although we rely on subsampling to enable the digitisation of high-frequency signals. One of the linearity limitations for successive-approximation A/D converters is the non-linearity of the DC reference generator (D/A converter). The reader is referred to Reference 28 for some guidelines on the software generation and optimisation of periodic noise-shaped bit sequences for DC level generation. As for the averaging filter that is used to extract the encoded DC levels, only a passive on-chip filter is implemented [28, 31], and the reason is to maximise robustness to process variations and to minimise design effort. For guidelines on the design of the passive filter, the reader is referred to Reference 32.

It should be noted that the digitisation algorithm described here relies primarily on the proper synchronisation with the excitation system. In each comparison pass, the comparator is expected to see the exact same samples of the test signal as it did in the previous passes. This is easily achieved using the bit stream generation approach, since all the generated DUT stimulus tone phases are well defined with respect to the sampling clock, F_S. On another note, the clock speed of the periodic bit stream generators is generally faster than the comparison speed of typical integrated analogue comparators. Under these circumstances, a sample-and-hold (S/H) circuit (which can track and sample very fast signals) can be inserted at the corresponding input of the comparator to receive the signal from the circuit under test. The addition of such a circuit allows the comparator to take multiple master clock cycles ($1/F_S$) to arrive at a decision about the relative magnitudes of its two input signals without compromising the effective sampling rate of the digitiser. The only requirement is that the ratio of F_S and the comparator clock frequency has to be a prime relative to the total number of samples in a test period, N, which is the same as the requirement for coherent sampling [28]. Similarly, when broadband signal digitisation is sought, the comparator clocking system can be modified further according to the methods in [28] to result in a high-bandwidth and flexible on-chip measurement solution.

Processing of the comparator output is done using a multibit memory (Figure 10.7) that is the same length as that of memory A and that is initialised to zero at the beginning of a measurement run. For each comparison pass, the bit stream generators continuously circulate their contents to output the analogue stimulus and the appropriate reference level, respectively. For each circuit response sample, the corresponding memory location is incremented or left unchanged, depending on whether the comparator output (for the current DC level) was 1 or 0. At the end of the digitisation process, each memory location contains an integer count representing the quantisation level for the corresponding sample (i.e. a thermometer code).

10.5.3 Self-checking

In order to increase the confidence in the on-chip test results, it is important for the test core to at least provide a way of checking whether it works or not. The number of checks that can be performed on the test system depends on how much time and effort the final system on chip (SOC) integrator is willing to spend. For example, referring to Figure 10.10a, the following simple check can locate catastrophic failures in the comparator. The reference input to the comparator is set to mid-rail, and the AC bit stream generator output is fed directly to the comparator input. As such, the comparator simply looks like a digital latch, so a bit-for-bit comparison between the input bit stream and the stream coming out of the comparator can be performed. Note that this is a test that can be performed completely in software using digital techniques. Moreover, this test automatically uncovers stuck-at faults in the circular memory structures.

The above tests could be sufficient for some applications, and the comparator can then be assumed to function properly and can be used to check other circuits or the circuit under test (CUT). Alternatively, more rigorous testing of the comparator resolving power can be done as follows. Another replica of the DC generator can be implemented in hardware and connected to the 'signal' terminal of the comparator, while the usual DC generator is connected to the 'reference' input. This way, slightly differing DC signals can be applied at different common-mode levels in order to verify (to the extent to which the DC sources are matched) the resolving power of the comparator under different signal conditions (Figure 10.10b). Replicating the DC generator seems to be feasible because it consists of a mostly digital implementation and thus scales favourably with technology. Again, this test can also be performed in software, and it provides for a quick and cheap relative measure of the comparator resolving power.

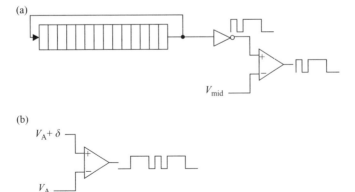

Figure 10.10   *(a) Converting the comparator into a digital latch to verify function-ality. (b) Replicating the DC generator allows for a quick method of sweeping comparator terminals*

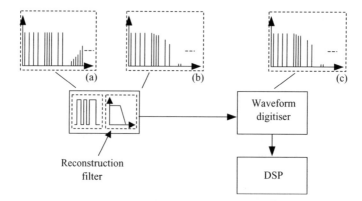

Figure 10.11   *Test core bypassing CUT in order to measure the analogue recon-*
struction filter (clocking system not shown for clarity). The response
(b) to a multitone signal (a) can be captured using an integrated
digitiser (c) in order to predict filter behaviour

If an analogue filter is implemented as part of the test core, then it too will
have to be characterised in order to maximise the available test signal bandwidth.
Referring to Figure 10.11, now that the comparator is verified, the implemented
filter magnitude and phase response can be measured using our integrated capture
system using a multitone stimulus signal [6]. The filter response can then be compensated for in the bit stream generation phase. Other types of tests can also be
performed on the filter before the overall test system can be used to verify the other
integrated analogue circuits. For example, in a manner similar to [33], a model
of the non-linearity introduced by the filter can be created, and its effects can be
cancelled in software when, say, the non-linearity of the circuit under test is being
verified.

10.6   Circuit considerations and instrument performance evaluation

In this section, we delve into some of the design considerations for the analogue
components of the test core described in Section 10.5 (comparator and sampler circuits). We do so in the context of an 8-bit integrated prototype that was fabricated
in a 3.3 V, 0.35 μm standard CMOS process. Although many prototypes have been
fabricated to demonstrate different aspects of the applicability of the proposed test
core to the embedded test of analogue circuits, this prototype is sufficient to raise most
of the design issues. In this section, we also present an experimental evaluation of the
raw performance of the on-chip instruments in this prototype. Just like external test
equipment, an evaluation of such performance capability is important. A description
of how the measurement instruments are applied to sample test cases is deferred until
Section 10.7.

10.6.1   *Circuit considerations*

The integrated prototype described here integrates all the components in Figure 10.7. Since the digital components can be automatically synthesised, and since the DC filter design can also be automated in software [32], we focus in this section on the voltage comparator and its associated S/H. A schematic of the comparator is illustrated in Figure 10.12 [34]. It consists of a preamplifier and a regenerative latch. During the track phase, the circuit behaves as a single-pole amplifier and boosts the input difference voltage slightly in order to provide a big enough seed for the latch. Once the latch signal is activated, diode-connected transistors M7/M8 are disconnected from the output nodes, and positive feedback due to cross-coupled transistors M9/M10 pushes the seed levels towards V_{DD} or ground depending on the polarity of the comparison result. The gain of the amplifier in track mode was designed to be about 10 V/V in order to sufficiently overcome the latch offset and provide adequate amplification within a reasonable clock rate. Several comments about this comparator design should be made. First, since the input test signal is only subjected to one comparator input capacitance node, a relatively large differential pair could be used without worrying about significantly loading the driving circuitry. A large input pair means a reduced input-referred noise of the preamplifier, and hence the whole comparator. Enlarging the input pair does not threaten the implementation area since the comparator occupies only about 1.5 per cent of the overall test core area in our 0.35 μm process. Second, this comparator has the advantage of a relatively fast conversion time due to the use of both linear and non-linear amplification in the preamplifier and latch, respectively.

In the comparator of Figure 10.12, we do not employ any offset compensation scheme. Thus, the common-mode dependence of offset voltage is expected to be

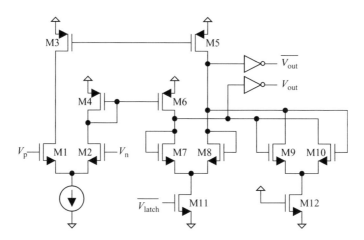

Figure 10.12   *Implemented comparator consisting of a single-pole low-gain linear amplification stage followed by a positive feedback (non-linear gain) latch [34]. No offset correction was necessary for target resolution*

relatively large. Such dependence would affect the linearity of the comparator and consequently the digitiser. In order to verify the viability of the comparator for the case of eight bits, a high-level MATLAB simulation of our digitisation algorithm (Figure 10.9) was performed, in which the common-mode dependence of the comparator offset was modelled. The conclusion drawn was that even an overly pessimistic signal-dependent offset variation was insignificant for the case of eight bits, so no offset cancellation was employed. This hypothesis has been verified experimentally as will be demonstrated shortly. It should also be noted that this discussion is a classic illustration of the design trade-offs involved in the implementation of the proposed test cores. The comparator is the only critical analogue component, and it should be kept as simple as could be for a given performance level. Fortunately, comparator design is governed by a small set of parameters when compared to other analogue building blocks, so its performance can be rapidly evaluated using high-level models.

An S/H circuit was placed in front of the comparator in order to provide a higher tracking bandwidth. Figure 10.13 shows a basic S/H topology that is expected to achieve very high-speed operation. When the switch (with on-resistance, R_{on}) is turned on, the effective bandwidth of this circuit is dependent on $R_{on}C_{hold}$, which could be made quite small. Complementary switch dimensions of $\sim 50\times$ the minimum width and a 300 fF hold capacitor achieve a bandwidth of about 4.5 GHz in a 0.35 μm CMOS technology. The variation of the on-resistance of the sampling switch is a major distortion mechanism at high frequencies, and it was minimised through careful sizing of the PMOS and NMOS devices. The only remaining consideration was to ensure that the clock is slow enough so that the first-order exponential settling transient has arrived to within the required amplitude resolution.

In order to combat the non-linearity of the S/H stage and to compensate for the signal-dependent charge injection (CI) of the switch, another replica of the S/H stage is placed at the negative comparator terminal. The reason is that we are only interested in the difference voltage across the two comparator inputs. The polarity of this difference

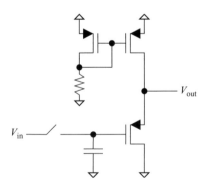

Figure 10.13   Implemented sample-and-hold amplifier. Tracking bandwidth is determined by the RC time constant of the switch on-resistance and the sampling capacitor

voltage should not be affected by variations in the S/H stage characteristic, as long as the latter is monotonic and as long as the two paths are properly matched. It has been suggested [35] that matching of up to 10 bits is possible with a topology that is very similar to the one adopted in this design.

10.6.2 Experimental evaluation

This section describes an evaluation of the on-chip instruments themselves. Thus, both the stimulus and digitisation functions of the prototype IC were evaluated by connecting the appropriate instruments to known calibrated external test equipment. For example, in order to evaluate the performance of the arbitrary waveform generator, several kinds of signals were generated and observed on bench instruments such as the HP3588A spectrum analyser. Here we show only one example of a bandpass signal in Figure 10.14. In this figure, a sine wave with a frequency of about 12 MHz is encoded in a periodic $\Sigma\Delta$ stream running at a clock frequency, F_S, of 250 MHz. The measured spectrum before filtering is shown in this figure, and the spectrum after (external) filtering using a passive filter is shown in Figure 10.15. As can be seen, with proper buffering and probing techniques, the encoded signal was extracted without significant degradation, even as the output samples were being updated at 4 ns intervals. It is interesting to note that the sine wave could have just as easily been placed at, say, 120 MHz with similar spectral purity. The reason is that spectral

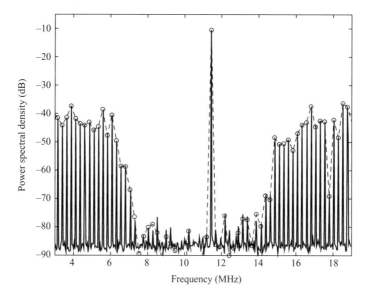

Figure 10.14 *Measured spectrum of a $\Sigma\Delta$ modulated stream at 250 MHz as seen on a HP3588A spectrum analyser. The dashed curve is an envelope of actual programmed frequency bin power in the software. Very few spurious tones (i.e. deviations from this envelope) are apparent in the measured result*

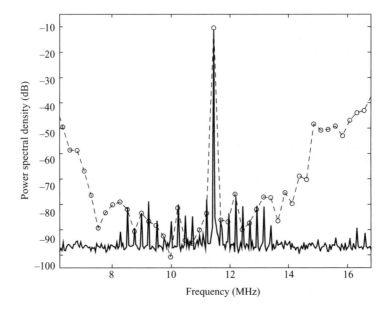

Figure 10.15   Measured spectrum of a $\Sigma\Delta$ stream at 250 MHz as seen on a HP3588A spectrum analyser. Filtering is performed externally using a discrete bandpass filter. The dashed curve is programmed tone power in software. Few spurious tones are apparent in the measured result

degradation is mainly due to the finite rise-time effects of the digital pulses, which are the same for a given clock frequency regardless of the location of the signal band.

The arbitrary waveform generator was then used to verify the performance of the integrated waveform digitiser. A single tone at the fundamental frequency of the system ($F_S/1024$) was generated at approximately 20 kHz ($F_S = 20$ MHz). This tone was fed directly to the digitiser, and the FFT of the digitised waveform is shown in Figure 10.16. In this and all subsequent results, it was ensured that spurs due to the AC source were well below the target performance level of the digitiser. A SFDR of more than 65 dB is achieved, which is consistent with the software predicted performance of the device.

Higher-frequency tests were performed in order to evaluate the viability of the test core for these frequencies. Although the AC bit stream generator was demonstrated to generate high-frequency bandpass signals, we did not have flexible high-frequency filtering capability with variable centre frequency to accommodate our bandpass $\Sigma\Delta$ modulated streams. An external RF source, an Agilent 83712B 20 GHz synthesised CW generator, was used instead. The AC bit stream generator was used to generate a 10 MHz digital synchronisation signal to which the CW generator locked on to. Synchronisation is achieved this way since our on-chip digitiser is driven by the same clock as that of the bit stream generator. Figure 10.17 shows the spectrum and reconstructed waveform of a 20.001 MHz signal that is digitised using our test core

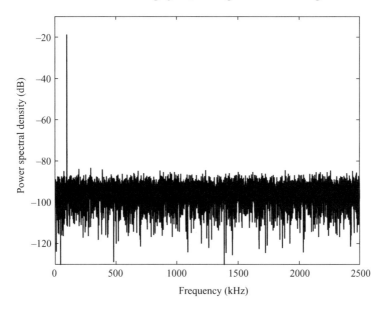

Figure 10.16   FFT-based spectral estimate of a digitised sine wave that is generated using the on-chip arbitrary waveform generator

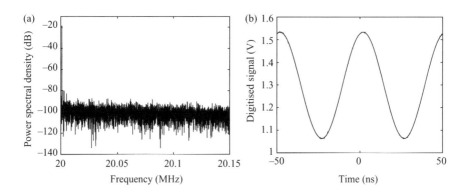

Figure 10.17   Sample result for a sine wave at 20.001 MHz digitised at a 20 MHz clock rate. (a) Power spectral density, (b) time-domain plot

at a rate of 20 MHz. As can be seen, a SFDR of about 61 dB is still achieved at this frequency.

Input stimuli at 200 MHz and higher were also digitised at a 20 MHz clock (always offset by 1 kHz), and the results are included in Figure 10.18. It should be noted that for these frequencies, an external resistor for wideband matching was placed at close proximity to the integrated prototype (which was housed in a ceramic flat pack surface-mount package capable of passing 4.5 GHz signals through short package leads)

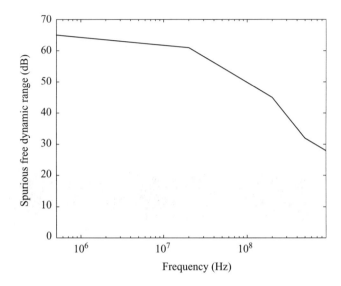

Figure 10.18   Digitiser performance as a function of test signal frequency. The observed performance is similar to some of the best published data converter performance in the same technology [36]

and connected to the source through a 50 Ω trace and cable. The degradation in linearity at these frequencies is attributed to clock jitter in our experimental setup. Specifically, the synchronisation signal travels through a long path from the chip to the CW generator, and it is thus corrupted by jitter. Since the generator is synchronised to this jittery signal, this will directly translate into spurious tones at its output. Despite these challenges, however, the results included in Figure 10.18 are almost on par with the best reported results in the literature on data converters targeting test applications and implemented in a similar technology [36]. Table 10.1 summarises the area and performance specifications of our integrated prototype.

10.7   Measurement types and test examples

In this section, we describe the kinds of tests that can be performed using the simple architecture of Figure 10.7. Specifically, this architecture can be programmed to perform DC curve tracing, oscilloscope, spectrum, and timing analysis functions while occupying an area equivalent to that of only a few thousand logic gates. The flexibility and compactness of such a solution makes it ideal for highly parallel testing at the block level within a complex IC or across different ICs that are tested simultaneously.

Table 10.1   Performance summary of integrated prototypes

Technology	3.3 V, 3 Metal, 0.35 μm CMOS
Amplitude resolution	8
Frequency resolution	10
Sampling rate	20 MHz (4 GHz effective)
Area	7K 2-input standard cell NAND gates (0.67 mm^2)
Analogue area, excluding DC RC area*	2.1% of Total (0.014 mm^2)
Analogue area, including DC RC area*	9% of Total (0.06 mm^2)
Maximum DNL/INL (@ 4 MHz digitiser clock)	0.15/0.4
SFDR @ F_{in} = 20.001 MHz	61 dB
THD @ F_{in} = 20.001 MHz	<0.15%
SFDR @ F_{in} = 200.001 MHz	45 dB
THD @ F_{in} = 200.001 MHz	0.7%
Maximum timing resolution	180 ps
Timing system RMS (P-P) jitter**	27 ps (264 ps)

* Area does not include AC filter, which could be part of the circuit under test
** Limited by experimental setup [28]

10.7.1   DC curve-tracer

DC characteristics often carry a lot of importance when characterising the large-signal behaviour of various components. Examples include the transfer characteristics of linear components and the DC linearity of A/D or D/A converters [6]. Our embedded test system can perform a DC sweep for these purposes. Specifically, the automatic waveform generator (AWG) is not limited to generating AC signals. It can be programmed to generate DC voltages using the periodic noise-shaped bit stream based generation technique described in this chapter. For each such voltage level, the DC output of the circuit under test can be digitised using our digitiser in much the same way that a voltmeter would perform a DC or RMS voltage measurement. Figure 10.19 shows such an experimental result where the DC transfer curve of a passive discrete RLC filter having a DC gain of about 0.57 V/V is measured. In this figure, the x-axis is the voltage that is encoded in our AWG, and the y-axis is the digitised DC voltage. As can be seen, the slope of the measured characteristic is about 0.57 V/V, so our test core is indeed able to extract the proper gain of the circuit.

10.7.2   Spectrum analyser

The presence of fully coherent stimulus and capture functions allows for great versatility in analogue testing. Much of the success of DSP-based techniques in reducing test time and enhancing accuracy is due to the power of coherent testing and the use of FFT techniques for frequency-domain analysis. In the proposed test core, the same versatility can be demonstrated because of the guaranteed coherence between signal

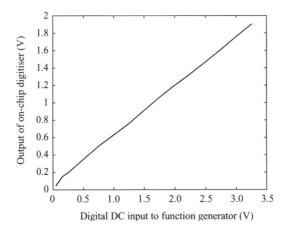

Figure 10.19 *Experimental result of a DC voltage sweep applied to an external RLC network*

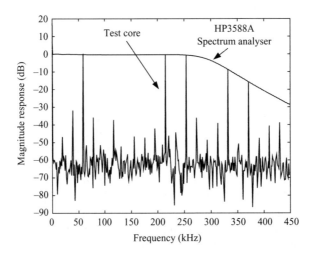

Figure 10.20 *Experimental result demonstrating frequency response measurement of a lowpass filter using multitone signals. A multitone signal containing five tones was generated using our test core, and the filter response to the tones was digitised and processed*

sources and digitisers. To illustrate the frequency-domain capability of our test core, consider the frequency response measurement depicted in Figure 10.20. The procedure for extracting the frequency response is as follows. First, a multitone signal is encoded in a periodic bit stream. If this bit stream is fed directly to the filter under test and the filter output digitised using the on-chip digitiser, a FFT-based analysis on the digitised samples can be performed. The example in Figure 10.20 is an experimental result using the integrated prototype of Section 10.6. The figure shows an overlay of the

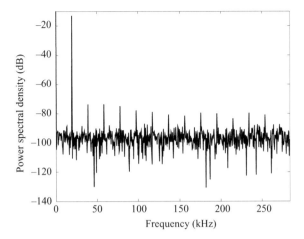

Figure 10.21   Spectral plot illustrating total harmonic distortion, SFDR, or SNR measurement

filter magnitude response as measured by a HP3588A spectrum analyser and the result of FFT analysis on the output of our on-chip waveform digitiser (sampling at a rate of 20 Msamples/s). The filter is an eighth-order Butterworth having a bandwidth of 300 kHz. The chosen multitone frequencies were at bins 3, 11, 13, 17, and 19, where bin 1 is at $F_S/1024$. It is important to note that the spurious tones are due to quantisation noise, and they can be suppressed if a higher quantiser granularity is chosen.

Without requiring a frequency sweep or multitone signals, other useful performance measures can be extracted using FFT-based analysis on our test system outputs. Some metrics such as SFDR, SNR, and harmonic distortion were described earlier, and they can be measured by passing only a single tone to the circuit under test and digitising its output. Figure 10.21 shows a sample experimental result of a harmonic distortion measurement. In this figure, a single tone with a peak-to-peak voltage of about 1.0 V was generated using our on-chip waveform generator. The sine wave was used to drive an external continuous-time filter, and the filter output was then digitised using our integrated prototype at a sampling rate of 20 MHz. FFT analysis was performed on the collected samples. From the figure, a total harmonic distortion of 0.19 per cent and a SFDR of about 61 dB are measured, and these numbers correlate well with an identical experiment that was performed using a HP33250A function generator with a HP3588A spectrum analyser.

10.7.3   Oscilloscope

In normal operation, the samples at the output of our on-chip waveform digitiser can be plotted directly to give a representation of the analogue signal under test. Figure 10.22 shows an experimental example of such an operation mode, in which a single tone is simply plotted in the time domain. This figure actually consists of an overlay of two plots: the output of our arbitrary waveform generator as seen on a

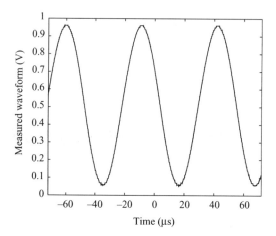

Figure 10.22   *Time-domain waveform illustrating an internally digitised on-chip sine wave. This and other measurements are useful not only for production purposes, but also for debug, diagnosis and design characterisation*

HP54602B digital oscilloscope, and the same signal after it was digitised using our on-chip waveform digitiser. As can be seen, the waveforms are indistinguishable. Having this ability to observe the steady-state or repetitive transient behaviour of signals internal to an IC is significant. It is not only useful for tests during production, but it is also invaluable for device characterisation or debug. Because of the complex modes of operation and the reliance on physical operating modes of active devices, analogue design is still (unfortunately) error-prone, and prototype evaluation and characterisation is an indispensable step in the process of design and manufacture. Such simple design validation capability as the one presented in Figure 10.22 can be a major aid in determining basic functionality. Many signal artefacts, such as clipping, excessive DC offset, or noise can be rapidly assessed through a simple observation of the internal time-domain waveforms.

As was mentioned in Section 10.5, a simple modification to the comparator clocking system allows the digitiser to capture wideband phenomena without altering the overall system architecture [28]. The following examples illustrate how this subsampling mechanism is used to capture signals running at speeds equal to or more than the sampling clock. Specifically, Figure 10.23 shows a reconstructed waveform of a 25 MHz digital clock, obtained at an effective sample resolution of 1.6 GHz. The waveform is progressively reconstructed using a subsampling digitiser clock that is also running at 25 MHz [28]. The way this delayed-clock subsampling digitisation is performed is as follows. First, zero phase delay between the test core's master clock and the digitiser clock is set, and the input signal is digitised according to the algorithm described in Section 10.5. Once all the signal samples under this phase condition are digitised and stored in memory, the digitiser clock is delayed by an amount equal to 625 ps with respect to the system clock, and the input signal is again

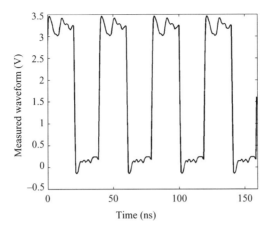

Figure 10.23   Captured rectangular waveform running at 25 MHz. Delayed-clock subsampling was used in order to achieve an effective sample rate of 1.6 GHz in a 0.35 μm CMOS process

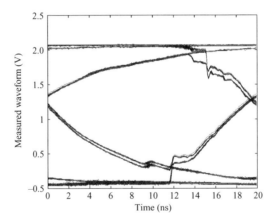

Figure 10.24   Illustration of an eye-diagram measurement. A long pseudo-random digital signal was passed through a lowpass filter to emulate a lossy channel. The output of the filter was sampled using the proposed test cores at a rate of 1.6 GHz

digitised. The process of incrementing the phase of the digitiser clock is repeated until the whole sampling clock period has been covered.

It is interesting to note that the broadband signal being measured can also be a long pseudo-random bit sequence (PRBS), which is useful for testing GHz-rate transceiver circuits. Such a sequence can be applied to a DUT with the response of the DUT captured using the above subsampling method. In this case, an eye diagram over a unit interval can be constructed, where certain eye (signal) templates are tested to ensure an appropriate error performance rate [37]. Figure 10.24 illustrates an example

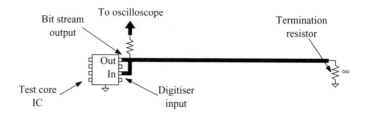

Figure 10.25   Experimental setup used to demonstrate socket/board test applications involving time-domain measurements. The probe connection to the oscilloscope was included to verify the setup

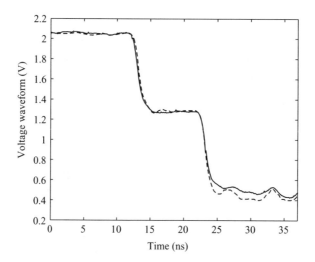

Figure 10.26   Experimental time-domain reflectometry measurement. The termination resistor in this case is an open-circuit. (-) Integrated prototype, (–) Tektronix TDS8000 digital sampling oscilloscope with 40 GHz sampling head

of such a measurement. In this figure, the sequence being measured is slow (25 MHz), and the eye opening is artificially reduced by passing the sequence through a lowpass filter in order to emulate channel losses and inter-symbol interference in a manner similar to Reference 38.

Before closing this section, a final useful oscilloscope function is demonstrated. Specifically, a time-domain reflectometry measurement at a sample resolution of 4 GHz is illustrated in Figures 10.25 and 10.26. Specifically, Figure 10.25 illustrates an experimental setup in which signal transmission and reflection on a piece of board interconnect is characterised using our integrated prototype. Figure 10.26 shows the actual measured result. As can be seen, a pulse was transmitted down the unterminated printed circuit board line, and the voltage at the input of the line was sampled and digitised using our integrated test core. As can be expected, time-domain measurements like this one are important in characterising the electrical characteristics of components such as packages and external interconnect discontinuities. Figure 10.26 also compares our result to that obtained by connecting the same test point to a Tektronix TDS8000 digital sampling oscilloscope using a carefully designed custom-built high-speed probe. Care was taken to ensure that the measured signal does not get altered much as it propagates through the probe and cables to the oscilloscope.

10.7.4 Jitter measurement device

If an edge-uncertainty or jitter measurement is sought, the above procedure to broadband signal capture can be used, except that it is not necessary to perform a full sweep of all possible quantisation levels in each digitisation pass. Specifically, for these tests, time instance measurements are being evaluated, and these are defined by the time a signal crosses a certain fixed threshold. Thus, the comparator reference can be held to a constant level. If this is the case, noise present on the input signal being measured will alter the time this signal crosses the comparator threshold. Referring to Figure 10.27, repetitively sampling a jittery digital edge with a fixed-threshold comparator would reveal that the edge transition will occur at different times, and the variability of the transition location is going to be determined by the amount of jitter present on the signal. Since we already have a means for shifting the sampling instance of our digitiser, an application of the delayed-clock subsampling mechanism described earlier can provide information about the jitter of a signal under test. All that

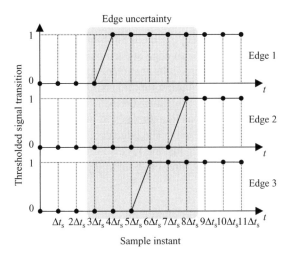

Figure 10.27   *Repetitively observing the time a signal crosses a fixed threshold reveals statistical information about the underlying variability in the signal period, or jitter. Many edges have to be captured and their relative frequency of occurrence over fixed sample instances compared*

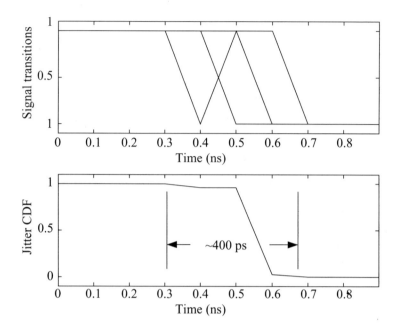

Figure 10.28   Sample jitter measurement. Edge occurrences at the different sampling instances are accumulated in order to obtain a representation of the jitter distribution function. In this example, sampling instances were spaced at 0.1 ns apart

needs to be performed is that many such edges of the repetitive signal are obtained, and the number of occurrences of an edge at each phase delay, Δt_S, is computed. By counting the occurrence of logical highs at each of the phase delays, an edge density graph or a jitter cumulative distribution function (CDF) can be obtained [38].

An experimental result of a jitter measurement is included in Figure 10.28. In this figure, several edge transitions like the ones in Figure 10.27 are shown overlaid, one on top of the other for simplicity. The edge density graph (the count of all edge transitions at each of the sampling instants) is also illustrated in the figure. As can be seen, in this case, a peak-to-peak jitter of about 400 ps has been measured.

10.8   Conclusions

In this chapter, some of the techniques, challenges and recent developments in the test and characterisation of RF, analogue, and mixed-signal circuits have been described. Specifically, present test techniques were introduced and shown to rely on direct access methods using external test equipment. These methods require the synchronisation of different pieces of measurement apparatus and rely on excessively long interconnection media for electronic signal delivery to/from the device being tested.

Such a measurement paradigm is becoming increasingly infeasible as modern mixed-signal devices continue to require tight measurement specifications and higher test signal bandwidths while offering a reduced number of interface pins for direct external access. Thus, although clever test list ordering and optimisation can be employed, some fundamental changes to test access methods have to be sought. To tackle this problem, this chapter introduced new techniques that can significantly enhance the testability and diagnostic ability of mixed-signal integrated circuits. At the heart of these techniques is a highly robust, compact, scalable and easily synthesisable embedded test core that emulates the functions of fully-fledged production-phase automatic test equipment. The integration of such a core represents a radical change from the present test practice; however, it is envisioned to provide a practical solution to an impending problem facing the semiconductor industry.

10.9   References

1 'The international technology roadmap for semiconductors' (Semiconductor Industry Association, 2001 Edition)
2 RABAEY, J.: 'Digital integrated circuits: a design perspective' (Englewood Cliffs, New Jersey: Prentice Hall, 1996)
3 ROBERTS, G. W.: 'Metrics, techniques, and recent developments in mixed-signal testing', Digest of technical papers. IEEE International Conference on *Computer-Aided Design*, San Jose, 1996, pp. 10–14
4 RAZAVI, B.: 'RF microelectronics' (Prentice Hall, Englewood Cliffs, New Jersey, 1998)
5 KENINGTON, P. B.: 'Emerging technologies for software radio', *IEE Electronics and Communication Engineering Journal*, April 1999, **11** (2), pp. 69–83
6 BURNS, M. and ROBERTS, G. W.: 'An introduction to mixed-signal IC test and measurement' (Oxford University Press, New York, 2001)
7 MILOR, L. S.: 'A tutorial introduction to research on analog and mixed-signal circuit testing', *IEEE Transactions on Circuits and Systems – II: Analog and Digital Signal Processing*, October 1998, **45** (10), pp. 1389–407
8 HUSS, S. D. and GYURCSIK, R. S.: 'Optimal ordering of analog integrated circuit tests to minimize test time'. Proceedings of IEEE Design Automation Conference, San Francisco, 1991, pp. 494–99
9 MILOR, L. and SANGIOVANNI-VINCENTELLI, A. L.: 'Minimizing production test time to detect faults in analog circuits'. *IEEE Transactions on Computer-Aided Design of Integrated Circuits*, June 1994, **13** (6), pp. 796–813
10 MILOR, L. and SANGIOVANNI-VINCENTELLI, A. L.: 'Optimal test set design for analog circuits'. Proceedings of IEEE International Conference on Computer-Aided Design, Santa Clara, 1990, pp. 294–97
11 SOUDERS, T. M. and STENBAKKEN, G. N.: 'Cutting the high cost of testing', *IEEE Spectrum*, March 1991, **28** (3), pp. 48–51

12 MALY, W., STROJWAS, A. J. and DIRECTOR, S. W.: 'VLSI yield prediction and estimation: a unified framework', *IEEE Transactions on Computer-Aided Design of Integrated Circuits*, January 1986, **5** (1), pp. 114–30

13 NAGI, N., CHATTERJEE, A. and ABRAHAM, J. A.: 'Fault simulation of linear analog circuits', *Journal of Electronic Testing: Theory and Applications*, 1993, **4**, pp. 345–60

14 SACHDEV, M. and ATZEMA, B.: 'Industrial relevance of analog IFA: a fact or a fiction'. Proceedings of IEEE International Test Conference, Washington, DC, October 1995, pp. 61–70

15 CHAO, C. Y., LIN, H. J. and MILOR, L.: 'Optimal testing of VLSI analog circuits', *IEEE Transactions on Computer-Aided Design of Integrated Circuits*, January 1997, **16** (1), pp. 58–77

16 TEMES, G. C.: 'Efficient methods of fault simulation'. Proceedings of IEEE Midwest Symposium on Circuits and Systems, Lubbot, TX, 1977, pp. 191–94

17 ZWOLINSKI, M., BROWN, A. D., CHALK, C. D.: 'Concurrent analog fault simulation'. Proceedings of IEEE International Mixed-Signal Test Workshop, Seattle, June 1997, pp. 42–47

18 TIAN, M. W. and SHI, R.: 'Efficient DC fault simulation of nonlinear analog circuits'. Proceedings of IEEE Design, Automation and Test in Europe Conference, Paris, 1998, pp. 899–904

19 HUYNH, S. D., KIM, S., SOMA, M. and ZHANG, J.: 'Automatic analog test signal generation using multifrequency analysis', *IEEE Transactions on Circuits and Systems – II: Analog and Digital Signal Processing*, **46** (5), May 1999, pp. 565–75

20 DEVARAYANADURG, G., SOMA, M., GOTETI, P. and HUYNH, S. D.: 'Test set selection for structural faults in analog IC's', *IEEE Transactions on Computer-Aided Design of Integrated Circuits and Systems*, July 1999, **18** (7), pp. 1026–38

21 GROCHOWSKI, A., BHATTACHARYA, D., VISWANATHAN, T. R. and LAKER, K.: 'Integrated circuit testing for quality assurance in manufacturing: history, current status, and future trends', *IEEE Transactions on Circuits and Systems – II: Analog and Digital Signal Processing*, August 1997, **44** (8), pp. 610–33

22 HELMREICH, K.: 'Test path simulation and characterization'. Proceedings of IEEE International Test Conference, Baltimore, MD, 2001, pp. 415–23

23 ROBERTS, G. W.: 'Improving the testability of mixed-signal integrated circuits'. Proceedings of IEEE Custom Integrated Circuits Conference, Santa Clara, 1997, pp. 214–21

24 HAWRYSH, E. M. and ROBERTS, G. W.: 'An integration of memory-based analog signal generation into current DFT architectures'. Proceedings of IEEE International Test Conference, Washington, DC, 1996, pp. 528–37

25 DUFORT, B. and ROBERTS, G. W.: 'On-chip analog signal generation for mixed-signal built-in self-test', *IEEE Journal of Solid-State Circuits*, March 1999, **34** (3), pp. 318–30

26 MAHONEY, M. V.: 'DSP-based testing of analog and mixed-signal circuits' (IEEE Computer Society Press, Washington, DC, 1987)

27 SCHREIER, R.: 'An empirical study of high-order single-bit delta-sigma modulators', *IEEE Transactions on Circuits and Systems – II: Analog and Digital Signal Processing*, August 1993, **40** (8), pp. 461–6

28 HAFED, M., ABASKHAROUN, N. and ROBERTS, G. W.: 'A 4-GHz effective sample-rate integrated test core for analog and mixed-signal circuits', *IEEE Journal of Solid-State Circuits*, April 2002, **37** (4), pp. 499–514

29 HAJJAR, A. and ROBERTS, G. W.: 'A high speed and area efficient on-chip analog waveform extractor'. Proceedings of IEEE International Test Conference, 1998, pp. 688–97

30 LOFSTROM, K.: 'Early capture for boundary scan timing measurements'. Proceedings of IEEE International Test Conference, Washington, DC, 1996, pp. 417–22

31 HAFED, M., LABERGE, S. and ROBERTS, G. W.: 'A robust deep submicron programmable DC voltage generator'. Proceedings of IEEE International Symposium on Circuits and Systems, Geneva, 2000, **4**, pp. 5–8

32 LABERGE, S.: 'DC voltage generation using periodic bit-stream modulation'. Masters Thesis, McGill University, 2002

33 HSIEH, L. and KUMAR, S. P.: 'Digitizer error extraction in the nonlinearity test'. Proceedings of IEEE International Test Conference, Washington, DC, 1994, pp. 757–62

34 SONG, B., LEE, S. and TOMPSETT, M.: 'A 10-b 15-MHz CMOS recycling two-step A/D converter', *IEEE Journal of Solid-State Circuits*, December 1990, **25** (6), pp. 1328–38

35 RAZAVI, B.: 'Principles of data conversion system design' (IEEE Press, Washington, DC, 1995)

36 POULTON, K., NEFF, R., MUTO, A., LIU, W., BURSTEIN, A. and HESHAMI, M.: 'A 4GSample/S 8b ADC in 0.35 μm CMOS', *Digest IEEE International Solid-State Circuits Conference*, San Francisco, 2002, pp. 166–67

37 KULP, B. D.: 'Testing and characterizing jitter in 100 Base-Tx and 155.52 Mbits/s ATM devices with 1 Gsample/s AWG in an ATE system'. Proceedings of IEEE International Test Conference, Washington, DC, 1996, pp. 104–11

38 ABASKHAROUN, N., HAFED, M. and ROBERTS, G. W.: 'Strategies for on-chip sub-nanosecond signal capture and timing measurement'. Proceedings of IEEE International Symposium on *Circuits and Systems*, Sydney, 2001, **4**, pp. 174–77

Index